Probability with Statistical Applications

Probability with Statistical Applications

Editor

Danyal Sadik

scitus academics

Probability with Statistical Applications
Edited by **Danyal Sadik**

ISBN: 9781681175645
Library of Congress Control Number: 2016937610

SCITUS Academics LLC,
www.scitusacademics.com
Box No. 4766, 616 Corporate Way,
Suite 2, Valley Cottage,
NY 10989

Notice

Reasonable efforts have been made to publish reliable data and views articulated in the chapters are those of the individual contributors, and not necessarily those of the editors or publishers. Editors or publishers are not responsible for the accuracy of the information in the published chapters or consequences of their use. The publisher believes no responsibility for any damage or grievance to the persons or property arising out of the use of any materials, instructions, methods or thoughts in the book. The editors and the publisher have attempted to trace the copyright holders of all material reproduced in this publication and apologize to copyright holders if permission has not been obtained. If any copyright holder has not been acknowledged, please write to us so we may rectify.

Printed in United States of America on Acid Free Paper

Preface

Probability is the measure of the likelihood that an event will occur. Probability is quantified as a number between 0 and 1 (where 0 indicates impossibility and 1 indicates certainty). The higher the probability of an event, the more certain we are that the event will occur. Randomness and uncertainty exist in our daily lives as well as in every discipline in science, engineering, and technology. Statistics and probability are sections of mathematics that deal with data collection and analysis. Probability is the study of chance and is a very fundamental subject that we apply in everyday living, while statistics is more concerned with how we handle data using different analysis techniques and collection methods. These two subjects always go hand in hand and thus you can't study one without studying the other. Probability theory is applied in everyday life in risk assessment and in trade on financial markets. Governments apply probabilistic methods in environmental regulation, where it is called pathway analysis. In addition to financial assessment, probability can be used to analyze trends in biology (e.g. disease spread) as well as ecology. As with finance, risk assessment can be used as a statistical tool to calculate the likelihood of undesirable events occurring and can assist with implementing protocols to avoid encountering such circumstances. Another significant application of probability theory in everyday life is reliability.

Probability with Statistical Applications features a wide range of important topics in modern probability theory and statistical applications. The book's coverage is thorough, its presentation logical and geared to student's needs. This book provides a versatile and lucid treatment of classic as well as modern probability theory, while integrating them with core topics in statistical applications.

Table of Contents

Chapter 1

The Discrete Poisson-Sujatha Distribution

Rama Shanker

Department of Statistics, Eritrea Institute of Technology, Asmara, Eritrea

ABSTRACT

In this paper, Poisson-Sujatha distribution has been obtained by compounding Poisson distribution with Sujatha distribution introduced by Shanker (2015 a). The first four moments about origin and the moments about mean has been obtained. The expression for coefficient of variation, skewness and kurtosis has been obtained. Important mathematical and statistical properties of the distribution have been derived and discussed. The estimation of its parameter has been discussed using both maximum likelihood estimation and the method of moments. The proposed distribution has been fitted using maximum likelihood estimate to some count data - sets to test its goodness of fit and the fit is compared with that obtained using other one parameter discrete distributions.

Keywords: Sujatha distribution, Compounding, Moments, Mathematical and statistical properties, Estimation of parameter, Goodness of fit.

1. INTRODUCTION

The probability density function (p.d.f) of one parameter Sujatha distribution introduced by Shanker (2015 a) for modeling real lifetime data sets from engineering and biomedical science is given by

$$f(x;\theta)=\frac{\theta^3}{\theta^2+\theta+2}\left(1+x+x^2\right)e^{-\theta x}\quad ;x>0,\ \theta>0 \tag{1.1}$$

Its corresponding cumulative distribution function (c.d.f) is given by

$$F(x)=1-\left[1+\frac{\theta x(\theta x+\theta+2)}{\theta^2+\theta+2}\right]e^{-\theta x};x>0,\theta>0 \tag{1.2}$$

It has been shown by Shanker (2015 a) that Sujatha distribution is a three component mixture of an exponential distribution with scale parameter θ , a gamma distribution having shape parameter 2 and a scale parameter θ , and a gamma distribution having shape parameter 3 and a scale parameter θ with their mixing proportions $\frac{\theta^2}{\theta^2+\theta+2}$, $\frac{\theta}{\theta^2+\theta+2}$ and $\frac{2}{\theta^2+\theta+2}$ respectively. Shanker (2015 a) has discussed its various mathematical and statistical properties including its shape, moment generating function, moments, skewness, kurtosis, hazard rate function, mean residual life function, stochastic orderings, mean deviations, distribution of order statistics, Bonferroni and Lorenz curves, Renyi entropy measure, stress-strength reliability , some amongst others. Further, Shanker (2015 a) has also discussed its estimation of parameter using maximum likelihood estimation and method of moments along with applications for modeling lifetime data and observed that it gives much closer fit than Akash and Shanker distributions introduced by Shanker (2015 b, 2015 c), Lindley (1958) and exponential distributions. It would be recalled that Shanker (2015 b, 2015 c) has proposed Akash and Shanker distributions along with their various interesting mathematical and statistical properties to model lifetime data arising from engineering and biomedical sciences and showed that these distributions provide much closer fit than Lindley and exponential distributions.

In the present paper, a Poisson mixture of Sujatha distribution introduced by Shanker (2015 a) has been obtained. The first four moments about origin and the moments about mean have been obtained and thus the expression for coefficient of variation, skewness and kurtosis has been given. The estimation of its parameter has been discussed using maximum likelihood estimation and method of moments. The distribution has been fitted using maximum likelihood estimate to some count data sets to test its goodness of fit over Poisson distribution and Poisson-Lindley distribution (PLD), a Poisson-mixture of Lindley distribution (1958), introduced by Sankaran (1970).

2. POISSON-SUJATHA DISTRIBUTION

Suppose the parameter λ of the Poisson distribution follows Sujatha distribution (1.1). Then the Poisson mixture of Sujatha distribution (1.1) can be obtained as

$$P(X=x)=\int_0^\infty \frac{e^{-\lambda}\lambda^x}{x!}\cdot\frac{\theta^3}{\theta^2+\theta+2}\left(1+\lambda+\lambda^2\right)e^{-\theta\lambda}d\lambda$$

$$=\frac{\theta^3}{\left(\theta^2+\theta+2\right)x!}\int_0^\infty \lambda^x\left(1+\lambda+\lambda^2\right)e^{-(\theta+1)\lambda}d\lambda \tag{2.1}$$

$$=\frac{\theta^3}{\theta^2+\theta+2}\cdot\frac{x^2+(\theta+4)x+\left(\theta^2+3\theta+4\right)}{(\theta+1)^{x+3}}\ ;x=0,1,2,...,\theta>0\,. \tag{2.2}$$

We name this distribution "Poisson-Sujatha distribution (PSD)". The graphs of the probability mass function (pmf) of PSD for varying values of its parameter θ are shown in the figure 1.

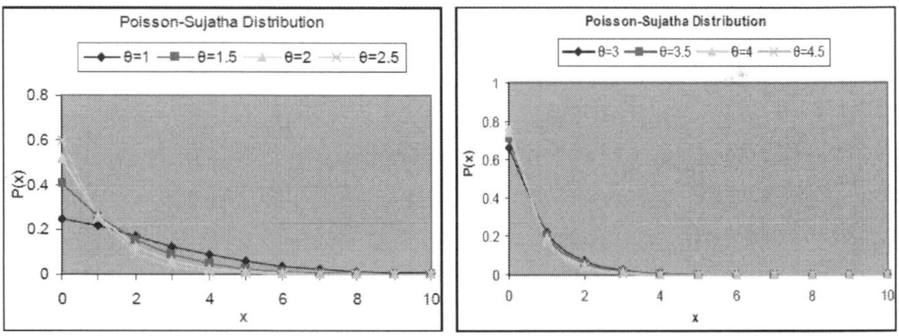

Figure 1. Graphs of probability mass function of PSD for varying val ues of the parameter θ

3. MOMENTS AND RELATED MEASURES

The rth factorial moment about origin of Poisson-Sujatha distribution (2.2) can be obtained as

$$\mu_{(r)}'=E\left[E\left(X^{(r)}\mid\lambda\right)\right], \text{where } X^{(r)}=X(X-1)(X-2)...(X-r+1)$$

Using (2.1) the rth moment about origin of PSD (2.2) can be obtained as

$$\mu_{(r)}'=E\left[E\left(X^{(r)}\mid\lambda\right)\right]=\frac{\theta^3}{\theta^2+\theta+2}\int_0^\infty\left[\sum_{x=0}^\infty x^{(r)}\frac{e^{-\lambda}\lambda^x}{x!}\right]\left(1+\lambda+\lambda^2\right)e^{-\theta\lambda}d\lambda$$

$$=\frac{\theta^3}{\theta^2+\theta+2}\int_0^\infty\left[\lambda^r\sum_{x=r}^\infty\frac{e^{-\lambda}\lambda^{x-r}}{(x-r)!}\right]\left(1+\lambda+\lambda^2\right)e^{-\theta\lambda}d\lambda$$

Taking $x+r$ in place of x within bracket, we get

$$\mu_{(r)}' = \frac{\theta^3}{\theta^2+\theta+2}\int_0^\infty \lambda^r \left[\sum_{x=0}^{\infty}\frac{e^{-\lambda}\lambda^x}{x!}\right]\left(1+\lambda+\lambda^2\right)e^{-\theta\lambda}\,d\lambda$$

The expression within the bracket is clearly unity and hence we have

$$\mu_{(r)}' = \frac{\theta^3}{\theta^2+\theta+2}\int_0^\infty \lambda^r\left(1+\lambda+\lambda^2\right)e^{-\theta\lambda}\,d\lambda$$

Using gamma integral and some algebraic simplification, we get finally a general expression for the rth factorial moment of PSD (2.2) as

$$\mu_{(r)}' = \frac{r!\left[\theta^2+(r+1)\theta+(r+1)(r+2)\right]}{\theta^r\left(\theta^2+\theta+2\right)}\ ;r=1,2,3,\ldots \qquad (3.1)$$

Substituting $r=1,2,3,\text{and }4$ in (3.1), the first four factorial moments can be obtained and using the relationship between factorial moments and moments about origin, the first four moments about origin of the PSD (2.2) are obtained as

$$\mu_1' = \frac{\theta^2+2\theta+6}{\theta\left(\theta^2+\theta+2\right)}$$

$$\mu_2' = \frac{\theta^3+4\theta^2+12\theta+24}{\theta^2\left(\theta^2+\theta+2\right)}$$

$$\mu_3' = \frac{\theta^4+8\theta^3+30\theta^2+96\theta+120}{\theta^3\left(\theta^2+\theta+2\right)}$$

$$\mu_4' = \frac{\theta^5+16\theta^4+84\theta^3+336\theta^2+840\theta+720}{\theta^4\left(\theta^2+\theta+2\right)}$$

Using the relationship between moments about mean and the moments about origin, the moments about mean of the PSD (2.2) are thus given by

$$\mu_2 = \sigma^2 = \frac{\theta^5+4\theta^4+14\theta^3+28\theta^2+24\theta+12}{\theta^2\left(\theta^2+\theta+2\right)^2}$$

$$\mu_3 = \frac{\theta^8+7\theta^7+32\theta^6+110\theta^5+228\theta^4+300\theta^3+240\theta^2+144\theta+48}{\theta^3\left(\theta^2+\theta+2\right)^3}$$

$$\mu_4 = \frac{\begin{pmatrix}\theta^{11}+15\theta^{10}+99\theta^9+488\theta^8+1682\theta^7+4016\theta^6+7008\theta^5+9016\theta^4\\ +8784\theta^3+6240\theta^2+2880\theta+720\end{pmatrix}}{\theta^4\left(\theta^2+\theta+2\right)^4}$$

The coefficient of variation $(C.V)$, coefficient of Skewness $(\sqrt{\beta_1})$, coefficient of Kurtosis (β_2) and index of dispersion (γ) of the PSD (2.2) are thus obtained as

$$C.V = \frac{\sigma}{\mu_1'} = \frac{\sqrt{\theta^5+4\theta^4+14\theta^3+28\theta^2+24\theta+12}}{\theta^2+2\theta+6}$$

$$\sqrt{\beta_1} = \frac{\mu_3}{\mu_2^{3/2}} = \frac{\theta^8+7\theta^7+32\theta^6+110\theta^5+228\theta^4+300\theta^3+240\theta^2+144\theta+48}{\left(\theta^5+4\theta^4+14\theta^3+28\theta^2+24\theta+12\right)^{3/2}}$$

$$\beta_2 = \frac{\mu_4}{\mu_2^2} = \frac{\begin{pmatrix}\theta^{11}+15\theta^{10}+99\theta^9+488\theta^8+1682\theta^7+4016\theta^6+7008\theta^5 \\ +9016\theta^4+8784\theta^3+6240\theta^2+2880\theta+720\end{pmatrix}}{\left(\theta^5+4\theta^4+14\theta^3+28\theta^2+24\theta+12\right)^2}$$

$$\gamma = \frac{\sigma^2}{\mu_1'} = \frac{\theta^5+4\theta^4+14\theta^3+28\theta^2+24\theta+12}{\theta\left(\theta^2+\theta+2\right)\left(\theta^2+2\theta+6\right)}$$

To study the nature and behavior of $\mu_1', \mu_2, \mathrm{C.V}, \sqrt{\beta_1}, \beta_2$ and γ of PSD (2.2), numerical values of these characteristics for varying values of the parameter θ have been presented in table 1. It is clear that μ_1', μ_2, and γ are decreasing whereas C.V., $\sqrt{\beta_1}$, β_2 are increasing for increasing values of the parameter θ.

Table 1.

	Values of θ for Poisson-Sujatha Distribution					
	1	2	3	4	5	6
μ_1'	2.25	0.875	0.5	0.340909	0.25625	0.204545
μ_2	5.1875	1.484375	0.726190	0.451963	0.320586	0.246039
CV	1.012270	1.392399	1.704336	1.972026	2.209573	2.425006
$\sqrt{\beta_1}$	1.467931	1.898599	2.205880	2.443548	2.646990	2.831109
β_2	5.944113	7.864044	9.479441	10.777366	11.910905	12.969309
γ	2.305555	1.696428	1.452381	1.325757	1.251067	1.202862

Graphs to study the nature and behavior of coefficient of variation, coefficient of skewness, and coefficient of kurtosis of PSD (2.2) for varying values of the parameter θ have been shown in the figure 2. It is obvious from the graphs that the coefficient of variation, coefficient of skewness, and coefficient of kurtosis of PSD are increasing for increasing values of the parameter θ.

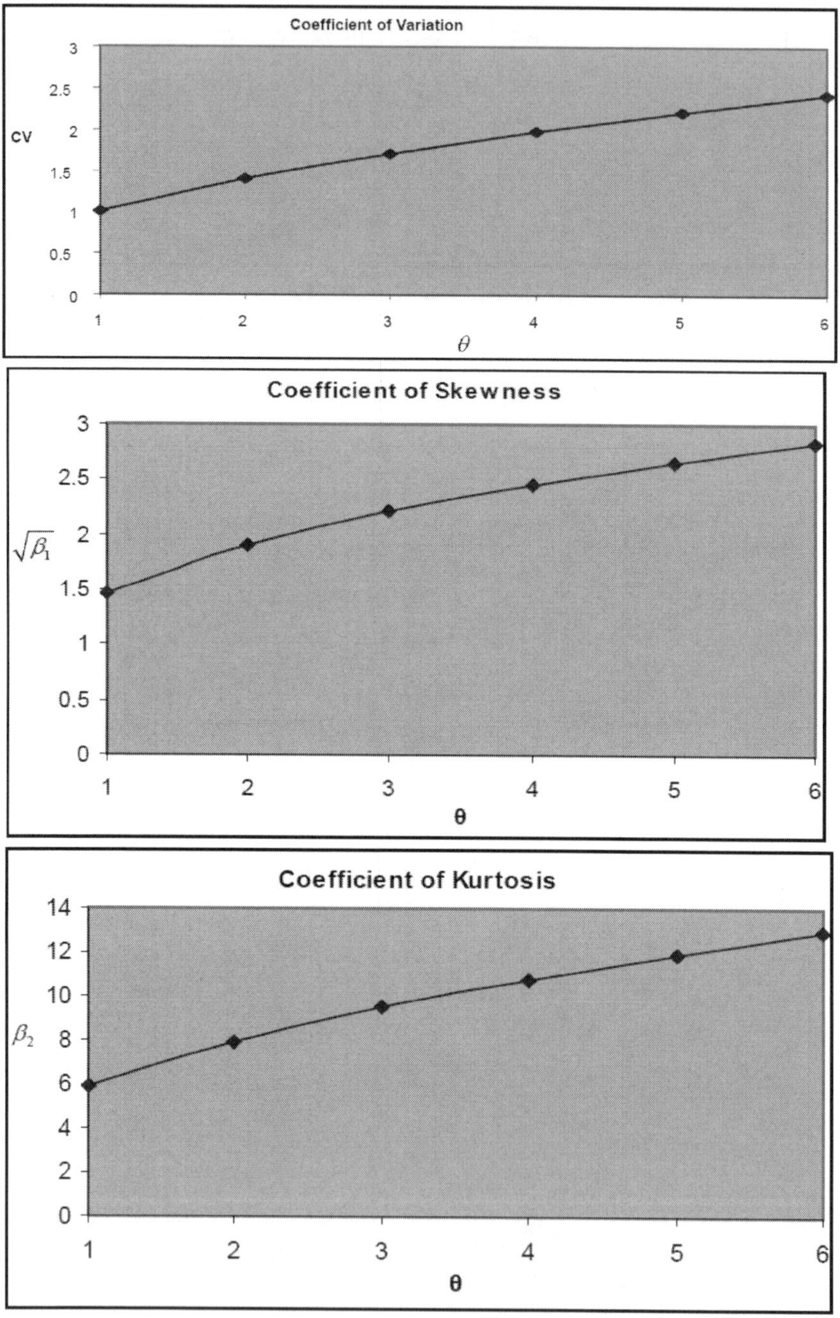

Figure 2. Graphs of coefficient of variation, coefficient of skewness and coefficient of kurtosis of PSD for varying values of the parameter θ

4. MATHEMATICAL AND STATISTICAL PROPERTIES

4.1. Over-Dispersion

The PSD (2.2) is always over- dispersed $(\sigma^2 > \mu)$. We have

$$\sigma^2 = \frac{\theta^5 + 4\theta^4 + 14\theta^3 + 28\theta^2 + 24\theta + 12}{\theta^2(\theta^2+\theta+2)^2}$$

$$= \frac{\theta^2 + 2\theta + 6}{\theta(\theta^2+\theta+2)}\left[\frac{\theta^5 + 4\theta^4 + 14\theta^3 + 28\theta^2 + 24\theta + 12}{\theta(\theta^2+\theta+2)(\theta^2+2\theta+6)}\right]$$

$$= \frac{\theta^2 + 2\theta + 6}{\theta(\theta^2+\theta+2)}\left[1 + \frac{\theta^4 + 4\theta^3 + 18\theta^2 + 12\theta + 12}{\theta(\theta^2+\theta+2)(\theta^2+2\theta+6)}\right]$$

$$= \mu\left[1 + \frac{\theta^4 + 4\theta^3 + 18\theta^2 + 12\theta + 12}{\theta(\theta^2+\theta+2)(\theta^2+2\theta+6)}\right] > \mu .$$

This shows that PSD (2.2) is always over-dispersed.

4.2. Increasing Hazard Rate (IHR) and Unimodality

The PSD (2.2) has an increasing hazard rate and unimodal. Since

$$\frac{P(x+1;\theta)}{P(x;\theta)} = \frac{1}{\theta+1}\left[\frac{(x+1)^2 + (\theta+4)(x+1) + (\theta^2+3\theta+4)}{x^2 + (\theta+4)x + (\theta^2+3\theta+4)}\right]$$

$$= \frac{1}{\theta+1}\left[1 + \frac{2x+\theta+5}{x^2 + (\theta+4)x + (\theta^2+3\theta+4)}\right]$$

is decreasing function in x, $P(x;\theta)$ is log-concave. Therefore, the PSD has an increasing hazard rate and unimodal. The interrelationship between log-concavity, unimodality and increasing hazard rate (IHR) of discrete distributions has been discussed in Grandell (1997).

4.3. Generating Functions

The probability generating function of the PSD (2.2) can be obtained as

$$P_X(t)=\frac{\theta^3}{(\theta^2+\theta+2)(\theta+1)^3}\left[\sum_{x=0}^{\infty}x^2\left(\frac{t}{\theta+1}\right)^x+(\theta+4)\sum_{x=0}^{\infty}x\left(\frac{t}{\theta+1}\right)^x+(\theta^2+3\theta+4)\sum_{x=0}^{\infty}\left(\frac{t}{\theta+1}\right)^x\right]$$

$$=\frac{\theta^3}{(\theta^2+\theta+2)(\theta+1)^3}\left[\frac{t(\theta+1)(\theta+1+t)}{(\theta+1-t)^3}+\frac{t(\theta+4)(\theta+1)}{(\theta+1-t)^2}\frac{(\theta^2+3\theta+4)(\theta+1)}{(\theta+1-t)}\right]$$

$$=\frac{\theta^3}{(\theta^2+\theta+2)(\theta+1)^2}\left[\frac{t(\theta+1+t)}{(\theta+1-t)^3}+\frac{t(\theta+4)}{(\theta+1-t)^2}\frac{(\theta^2+3\theta+4)}{(\theta+1-t)}\right]$$

The moment generating function of the PSD (2.2) is thus obtained as

$$M_X(t)=\frac{\theta^3}{(\theta^2+\theta+2)(\theta+1)^2}\left[\frac{e^t(\theta+1+e^t)}{(\theta+1-e^t)^3}+\frac{e^t(\theta+4)}{(\theta+1-e^t)^2}\frac{(\theta^2+3\theta+4)}{(\theta+1-e^t)}\right]$$

5. ESTIMATION OF THE PARAMETER

5.1. Maximum Likelihood Estimate (MLE) of the Parameter

Let $(x_1, x_2, \ldots, x_n)$ be a random sample of size n from the PSD (2.2) and let f_x be the observed frequency in the sample corresponding to $X=x\ (x=1,2,3,\ldots,k)$ such that $\sum_{x=1}^{k} f_x = n,$ where k is the largest observed value having non-zero frequency. The likelihood function L of the PSD (2.2) is given by

$$L=\left(\frac{\theta^3}{\theta^2+\theta+2}\right)^n\frac{1}{(\theta+1)^{\sum_{x=1}^{k} f_x(x+3)}}\prod_{x=1}^{k}\left[x^2+(\theta+4)x+(\theta^2+3\theta+4)\right]^{f_x}$$

The log likelihood function is thus obtained as

$$\log L=n\log\left(\frac{\theta^3}{\theta^2+\theta+2}\right)-\sum_{x=1}^{k} f_x(x+3)\log(\theta+1)+\sum_{x=1}^{k} f_x\log\left[x^2+(\theta+4)x+(\theta^2+3\theta+4)\right]$$

The first derivative of the log likelihood function is given by

$$\frac{d\log L}{d\theta}=\frac{n(\theta^2+2\theta+6)}{\theta(\theta^2+\theta+2)}-\frac{n(\bar{x}+3)}{\theta+1}+\sum_{x=1}^{k}\frac{\left[x+(2\theta+3)\right]f_x}{\left[x^2+(\theta+4)x+(\theta^2+3\theta+4)\right]}$$

where $\bar{x}$ is the sample mean.

The maximum likelihood estimate (MLE), $\hat{\theta}$ of θ of PSD (2.2) is the solution of the equation $\frac{d\log L}{d\theta}=0$ and is given by the solution of the non-linear equation

$$\frac{n\left(\theta^2+2\theta+6\right)}{\theta\left(\theta^2+\theta+2\right)}-\frac{n(\bar{x}+3)}{\theta+1}+\sum_{x=1}^{k}\frac{\left[x+(2\theta+3)\right]f_x}{\left[x^2+(\theta+4)x+\left(\theta^2+3\theta+4\right)\right]}=0$$

This non-linear equation can be solved by any numerical iteration methods such as Newton-Raphson, Bisection method, Regula – Falsi method etc.

5.2. Method of Moment Estimate (MOME) of the Parameter

Let $(x_1, x_2, \ldots, x_n)$ be a random sample of size n from the PSD (2.2). Equating the first moment about origin to the corresponding sample moment, the MOME $\tilde{\theta}$ of θ of PSD (2.2) is the solution of the following cubic equation $\bar{x}\theta^3+(\bar{x}-1)\theta^2+2(\bar{x}-1)\theta-6=0$ where $\bar{x}$ is the sample mean.

6. APPLICATIONS AND GOODNESS OF FIT

The PSD has been fitted to a number of count data - sets to test its goodness of fit over Poisson distribution (PD) and Poisson-Lindley distribution (PLD. The maximum likelihood estimate (MLE) has been used to fit the PSD. Five observed data-sets, for which the PD, PLD and PSD has been fitted, are presented. The first data-set is due to Kemp and kemp (1965) regarding the distribution of mistakes in copying groups of random digits, the second data-set is due to Beall (1940) regarding the distribution of Pyrausta nublilalis, the third data-set is the number of accidents to 647 women working on high explosive shells in 5 weeks, available in Sankaran (1970), the fourth and fifth data-sets are relating to the number of outbreaks of strike in U.K during 1948-1959 available in Consul (1989).

It is obvious from the fitting of PD, PLD, and PSD that the PSD gives much closer fit than PD and PLD in tables 1,2,3, and 5, whereas in table 4 PLD gives better fit than PD and PSD. It is to be noted that the fitting of PSD (2.2) in table 1 is also better than the generalized Poisson-Lindley distribution of Mahmoudi and Zakerzadeh (2010). Further, the fitting of PSD in table 3 is better than the fitting of negative binomial distribution (NBD) and the generalized Poisson-Lindley distribution of Mahmoudi and Zakerzadeh (2010).

Table 1. Distribution of mistakes in copying groups of random digits

No. of errors per group	Observed Frequency	Expected Frequency		
		PD	PLD	PSD
0	35	27.4	33.0	32.9
1	11	21.5	15.3	15.3
2	8	8.4	6.8	6.8
3	4	2.2	2.9	2.9
4	2	0.5	2.0	2.1
Total	60	60.0	60.0	60.0
ML estimate		$\hat{\theta} = 0.7833$	$\hat{\theta} = 1.7434$	$\hat{\theta} = 2.167811$
χ^2		7.98	2.20	1.63
d.f.		1	1	2
p-value		0.0047	0.1380	0.4426

Table 2. Distribution of Pyrausta nublilalis in 1937

No. of insects	Observed Frequency	Expected Frequency		
		PD	PLD	PSD
0	33	26.4	31.5	31.5
1	12	19.8	14.2	14.2
2	6	7.4	6.1	6.1
3	3	1.8	2.5	2.6
4	1	0.3	1.0	1.0
5	1	0.3	0.7	0.6
Total	56	56.0	56.0	56.0
ML estimate		$\hat{\theta} = 0.7500$	$\hat{\theta} = 1.8081$	$\hat{\theta} = 2.241454$
χ^2		4.87	0.53	0.45
d.f.		1	1	1
p-value		0.0273	0.4666	0.5023

Table 3. Accidents to 647 women working on high explosive shells

No. of accidents	Observed Frequency	Expected Frequency		
		PD	PLD	PSD
0	447	406	439.5	439.8
1	132	189	142.8	142.1
2	42	45	45.0	45.0
3	21	7	13.9	13.9
4	3	1	4.2	4.2
≥ 5	2	0.1	1.3	2.0
Total	647	647.0	647.0	647.0
ML estimate		$\hat{\theta} = 0.465$	$\hat{\theta} = 2.729$	$\hat{\theta} = 3.168063$
χ^2		61.08	4.82	2.75
d.f.		1	3	3
p-value		0.0273	0.1855	0.4318

Table 4. Number of outbreaks of strike in U.K during 1948-1959, Consul (1989)

No. of outbreaks of strikes	Observed Frequency	Expected Frequency		
		PD	PLD	PSD
0	110	103.5	110.2	110.3
1	33	42.5	32.7	32.5
2	9	8.7	9.4	9.4
3	3	1.2	2.7	2.7
4	1	0.1	1.0	1.1
Total	156	156.0	156.0	156.0
ML estimate		$\hat{\theta} = 0.410256$	$\hat{\theta} = 3.039932$	$\hat{\theta} = 3.473748$
χ^2		3.431	0.001	0.01
d.f.		1	1	1
p-value		0.0640	0.9748	0.9203

Table 5. Number of outbreaks of strike in U.K during 1948-1959, Consul (1989)

No. of outbreaks of strikes	Observed Frequency	Expected Frequency		
		PD	PLD	PSD
0	117	112.5	117.3	117.4
1	29	36.8	29.3	29.1
2	9	6.0	7.1	7.2
3	0	0.6	1.7	1.7
4	1	0.1	0.6	0.6
Total	156.0	156.0	156.0	156.0
ML estimate		$\hat{\theta} = 0.326923$	$\hat{\theta} = 3.801247$	$\hat{\theta} = 4.128452$
χ^2		3.458	0.04	0.03
d.f.		1	1	1
p-value		0.0629	0.8415	0.8625

7. CONCLUSIONS

In the present paper Poisson-Sujatha distribution (PSD) has been introduced by compounding Poisson distribution with Sujatha distribution proposed by Shanker (2015 a). Expression for *r*th factorial moment has been derived and the first four moments about origin and the moments about mean has been obtained. The expression for coefficient of variation, skewness and kurtosis has also been given. The estimation of its parameter has been discussed using maximum likelihood estimation and the method of moments. The PSD has been fitted using maximum likelihood estimate to some count data - sets to test its goodness of fit over Poisson distribution (PD) and Poisson-Lindley distribution (PLD) and it is found that Poisson-Sujatha distribution (PSD) is better than Poisson and Poisson-Lindley distributions for most of the count data-sets.

ACKNOWLEDGMENTS

The author is grateful for the reviewer's constructive comments.

REFERENCES

1. Beall, G. (1940): The fit and significance of contagious distributions when applied to observations on larval insects, *Ecology,* 21, 460-474.
2. Consul, P.C. (1989): Generalized Poisson Distributions, Dekker, New York.
3. Grandell, J. (1997): Mixed Poisson Processes, Chapman & Hall, London.
4. Kemp, C.D. and Kemp, A.W. (1965): Some properties of the Hermite distribution, *Biometrika*, 52, 381-394.
5. Lindley, D.V. (1958): Fiducial distributions and Bayes theorem, *Journal of the Royal Statistical Society*, 20 (1), 102-107.
6. Mahmoudi E. and Zakerzadeh, H. (2010): Generalized Poisson-Lindley distribution, *Communications in Statistics- Theory and Methods*, 39, 1785 – 1798.
7. Sankaran, M. (1970): The discrete Poisson-Lindley distribution, *Biometrics*, 26, 145-149.
8. Shanker, R. (2015 a): Sujatha distribution and Its Applications, Accepted for publication in *"Statistics in Transition new Series"*.

9. Shanker, R. (2015 b): Akash distribution and Its Applications, *International Journal of Probability and Statistics*, 4(3), 65 – 75.

10. Shanker, R. (2015 c): Shanker distribution and Its Applications, *International Journal of Statistics and Applications*, 5(6), 338 – 348.

Chapter 2

Bayesian Estimation and Inference for the Generalized Partial Linear Model

Haitham M. Yousof[1], **Ahmed M. Gad**[2]

[1]*Department of Statistics, Mathematics and Insurance, Benha University, Egypt*

[2]*Department of Statistics, Cairo University, Egypt*

ABSTRACT

In this article we propose a Bayesian regression model called the Bayesian generalized partial linear model which extends the generalized partial linear model. We consider Bayesian estimation and inference of parameters for the generalized partial linear model (GPLM) using some multivariate conjugate prior distributions under the square error loss function. We propose an algorithm for estimating the GPLM parameters using Bayesian theorem in more detail. Finally, comparisons are made between the GPLM estimators using Bayesian approach and the classical approach via a simulation study.

Keywords: Generalized Partial Linear Model, Profile Likelihood Method, Generalized Speckman Method, Back-Fitting Method, Bayesian Estimation

1. INTRODUCTION

The semi-parametric regression models are intermediate step between the fully parametric and nonparametric models. Many definitions of semi-parametric models are available in literature. The definition that will be adopted in this article is that the model is a semi-parametric model if it contains a nonparametric component in addition to a parametric component and they need to be estimated. The semi-parametric models are characterized by a finite-dimensional component, $\theta \epsilon \Theta \subseteq R^k$ and an infinite-dimensional $\gamma \epsilon \Gamma$.

Semi-parametric models try to combine the flexibility of a nonparametric model with the advantages of a parametric model. A fully nonparametric model will be more robust than semi-parametric and parametric models since it does not suffer from the risk of misspecification. On the other hand, nonparametric estimators suffer from low convergence rates, which deteriorate when considering higher order derivatives and multidimensional random variables. In contrast, the parametric model carries a risk of misspecification but if it is correctly specified it will normally enjoy $\sqrt{n}$-consistency with no deterioration caused by derivatives and multivariate data. The basic idea of a semi-parametric model is to take the best of both models. The semi parametric generalized linear model known as the generalized partial linear model (GPLM) is one of the semi-parametric regression models, See Powell (1994); Rupport et al. (2003); and Sperlich et al. (2006).

Many authors have tried to introduce new algorithms for estimating the semi-parametric regression models. Meyer et al. (2011) have introduced Bayesian estimation and inference for generalized partial linear models using shape-restricted splines. Zhang et al. (2014) have studied estimation and variable selection in partial linear single index models with error-prone linear covariates. Guo et al. (2015) have studied the empirical likelihood for single index model with missing covariates at random. Bouaziz et al. (2015) have studied semi-parametric inference for the recurrent events process by means of a single-index model.

The curse of dimensionality problem (COD) associated with nonparametric density and conditional mean function makes the nonparametric methods impractical in applications with many regressors and modest size samples. This problem limits the ability to examine data in a very flexible way for higher dimensional problems. As a result, the need for other methods became important. It is shown that semi parametric regression models can be of substantial value in solution of such complex problems.

Bayesian methods provide a joint posterior distribution for the parameters and hence allow for inference through various sampling methods. A number of methods for Bayesian monotone regression have been developed. Ramgopal et al (1993) introduced a Bayesian monotone regression approach using Dirichlet process priors. Perron and Mengersen (2001) proposed a mixture of triangular distributions where the dimension is estimated as part of the Bayesian analysis. Both Holmes and Heard (2003) and Wang (2008) model functions where the knot locations are free parameters with the former using a piecewise constant model and the latter imposing the monotone shape restriction using cubic splines and second-order cone programming with a truncated normal prior. Johnson (2007) estimates item response functions with free-knot regression splines restricted to be monotone by requiring spline coefficients to monotonically increasing. Neelon and Dunson (2004) proposed a piecewise linear model where the monotonicity is enforced via prior distributions; their model allows for flat spots in the regression function by using a prior that is a mixture of a continuous distribution and point mass at the origin. Bornkamp and Ickstadt (2009) applied their Bayesian monotonic regression model to dose-response curves. Lang and Brezger (2004) introduced Bayesian penalized splines for the additive regression model and Brezger and Steiner (2008) applied the Bayesian penalized splines model to monotone regression by imposing linear inequality constraints via truncated normal priors on the basis function coefficients to ensure monotonicity. Shively et al (2009) proposed two Bayesian approaches to monotone function estimation with one involving piecewise linear approximation and a Wiener process prior and the other involving regression spline estimation and a prior that is a mixture distribution of constrained normal distributions for the regression coefficients.

In this article, we propose a new method for estimating the GPLM based on Bayesian theorem using a new algorithm for estimation. The rest of the paper is organized as follows. In Section 2, we define the Generalized Partial Linear Model (GPLM). In Section 3, we present the Bayesian estimation and inference for the (GPLM). In Section 4, we provide Simulation Study. Finally, some concluding remarks and discussion are presented in Section 5.

2. GENERALIZED PARTIAL LINEAR MODEL (GPLM)

The GPLM model has the form

$$E[Y|X,W] = G[X\beta + m(W)], \tag{1}$$

where G(.) is a known link function. It is a semi-parametric model since it contains both parametric and nonparametric components. This model can be reduced to the following model,

$$E[Y|X,W] = X\beta + m(W). \tag{2}$$

For a unit linear function. It is called partial linear model (PLM).

The GPLM is used by Severini and Staniswalis (1994); Chen (1995); Burda et al. (1998); Muller (2001); Peng (2003); and Peng and Wang (2004).

The estimation methods for the GPLM, in Eq. (1), are based on the idea that an estimator $\widehat{\beta}$ can be found for known $m(.)$, and an estimator $\widehat{m}(.)$ can be found for known β. The estimation methods that will be considered are based on kernel smoothing methods in the estimation of the nonparametric component of the model, therefore the following estimation methods are presented in sequel.

2.1. Profile Likelihood Method

The profile likelihood method introduced by Severini and Wong (1992). It is based on assuming a parametric model for the conditional distribution of *Y* given *X* and *W*. The idea of this method is as follows:

First: Assume the parametric component of the model, i.e., the parameters vector, β.

Second: Estimate the nonparametric component of the model which depends on this fixed β, i.e. $m_\beta(.)$ by some type of smoothing method to obtain the estimator $\widehat{m}_\beta(.)$.

Third: Use the estimator $\widehat{m}_\beta(.)$ to construct profile likelihood for the parametric component using either a true likelihood or quasi-likelihood function.

Fourth: The profile likelihood function is then used to obtain an estimator of the parametric component of the model using a maximum likelihood method.

Thus the profile likelihood method aims to separate the estimation problem into two parts, the parametric part which is estimated by a parametric method and the nonparametric part which is estimated by a nonparametric method.

Murphy and Vaart (2000) showed that the full likelihood method fails in semi-parametric models. In semi parametric models the observed information, if it exits, would be an infinite-dimensional operator. They used profile likelihood rather than a full likelihood to overcome the problem, the algorithm for profile likelihood method is derived as follows:

Derivation of different likelihood functions

For the parametric component of the model, the objective function is the parametric profile likelihood function which is maximized to obtain an estimator for β is given by

$$\ell(\beta) = \sum_{i=1}^{n} \ell(\mu_{i,\beta}, y_i) = \sum_{i=1}^{n} \ell_i, \quad (3)$$

where $\ell(.)$ denotes the log-likelihood or quasi-likelihood function, $\mu_{i,\beta} = G[X_i^T\beta + m_\beta(W_i)]$, and $\ell_i = \ell(\mu_{i,\beta}, y_i)$.

For the nonparametric component of the model, the objective function is a smoothed or a local likelihood function which is given by

$$\ell_H[m_\beta(w)] = \sum_{i=1}^{n} \ell(\mu_{i,\beta(w)}, y_i)\, K_H(w - W_i), \tag{4}$$

where $\mu_{i,\beta(w)} = G[X_i^T\beta + m_\beta(W)] = G(\eta)$, and the local weight $K_H(w - W_i)$ is the kernel weight with $\mathcal{K}$ denoting a multidimensional kernel function and H is a bandwidth matrix. The function in Eq. (4) is maximized to obtain an estimator for the smooth function $m_\beta(w)$ at a point w.

Maximization of the likelihood functions

The maximization of the local likelihood in Eq. (4) requires solving

$$0 = \sum_{i-1}^{n} \dot{\ell}(G(\eta), y_i) K_H(w - W_i), \tag{5}$$

with respect to $\underline{m_\beta}(t)$. Note that $\dot{\ell}(.)$ denotes the first derivative of the likelihood function $\ell(.)$.

The maximization of the profile likelihood in Eq. (3) requires solving

$$0 = \sum_{\underline{i}\,1}^{n} \dot{\ell}(G(\eta), y_i) \left(U_i + \nabla_{m_\beta}(W_i)\right), \tag{6}$$

with respect to the coefficient vector β. The vector $\nabla_{\underline{m_\beta}}(W_i)$ denotes the vector of all partial derivatives of m_β with respect to β. A further differentiation of Eq. (5) with respect to *p* leads to an explicit expression for ∇_{m_β} as follows:

$$\nabla_{\underline{m_\beta}}(w) = -\frac{\sum_{i=1}^{n} \ddot{\ell}(G(\eta), y_i) K_H(w - W_i) U_i}{\sum_{i=1}^{n} \ddot{\ell}(G(\eta), y_i) K_H(w - W_i)}, \tag{7}$$

where $\ddot{\ell}$ denotes the second derivative of $\ell(G(\eta), y_i)$.

Equations (5) and (6) can only be solved iteratively. Severini and Saitniswalis (1994) presented a Newton-Raphson type algorithm for this problem as follows:

Let

$$\mathbf{m}_j = \mathbf{m}_\beta(\mathbf{W}_j),$$
$$\eta_i = (\mathbf{X}_i^T \boldsymbol{\beta} + \mathbf{m}_i),$$
$$\eta_{ij} = (\mathbf{X}_i^T \boldsymbol{\beta} + \mathbf{m}_j),$$
$$\ell_i = \ell(G(\eta_i), y_i),$$
$$\ell_{ij} = \ell(G(\eta_{ij}), y_i).$$

Further let $\dot{\ell}_i$, $\ddot{\ell}_i$, $\dot{\ell}_{ij}$ and $\ddot{\ell}_{ij}$ will be the first and second derivatives of ℓ_i and ℓ_{ij} with respect to their first argument. All values will be calculated at the observations W_i instead of the free parameter w. Then Equations (5) and (6) are transformed to

$$0 = \sum_{i=1}^{n} \dot{\ell}_{ij} \mathcal{K}_H (\mathbf{W}_i - \mathbf{W}_j), \tag{8}$$

and

$$0 = \sum_{i=1}^{n} \dot{\ell}_i (\mathbf{X}_i - \nabla_{m_i}), \tag{9}$$

respectively.

The estimator of $\nabla_{m_j} = \nabla_{m_\beta}(W_j)$ based on Eq. (7) is necessary to estimate $\boldsymbol{\beta}$

$$\nabla_{m_j} = -\frac{\sum_{i=1}^{n} \ddot{\ell}_{ij} \mathcal{K}_H(W_i - W_j) U_i}{\sum_{i=1}^{n} \ddot{\ell}_{ij} \mathcal{K}_H(W_i - W_j)}. \tag{10}$$

Equations (8) and (10) imply the following iterative Newton-Raphson Type algorithm

First: Initial values

Different strategies are found for obtaining start values of the estimators:

- Using $\hat{\boldsymbol{\beta}}^{(o)}$ and $\hat{\mathbf{m}}_j^{(o)}$ obtained from fitting a parametric generalized linear model (GLM).

- Using $\boldsymbol{\beta} = \mathbf{0}$ and $m_j = G^{-1}[(Y_j + \bar{Y})/2]$ with the adjustment for Binomial responses as $m_j = G^{-1}[(Y_j + 0.5)/2]$.
- Using $\boldsymbol{\beta} = \mathbf{0}$ and $m_j = G^{-1}[Y_j]$ with the adjustment for binomial responses as $m_j = G^{-1}[(Y_j + 0.5)/2]$. . (See Severini and Staniswalis, 1994).

Second: The updating step for $\boldsymbol{\beta}$

$$\boldsymbol{\beta}^{new} = \boldsymbol{\beta} - \mathbf{B}^{-1} \sum_{i=1}^{n} \dot{\ell}_i \widetilde{\mathbf{X}}_i \ ,$$

where **B** is a Hessian type matrix defined as

$$\boldsymbol{\beta} = \sum_{i=1}^{n} \ddot{\ell}_i \widetilde{\mathbf{X}}_i \widetilde{\mathbf{X}}_i^T \ ,$$

and

$$\widetilde{X}_j = X_j + \nabla_{m_j} = X_j - \frac{\sum_{i=1}^{n} \ddot{\ell}_{ij} \mathcal{K}_H (W_i - W_j) X_i}{\sum_{i=1}^{n} \ddot{\ell}_{ij} \mathcal{K}_H (W_i - W_j)}$$

$j = 1, \dots n$

The updating step for p can be summarized in a closed matrix form as follows:

$$\boldsymbol{\beta}^{new} = \left(\widetilde{X}^T D \widetilde{X}\right)^{-1} \widetilde{X}^T D \widetilde{Z},$$

where

$$\widetilde{X} = (1 - S^p)U,$$

$$\widetilde{Z} = \widetilde{X}\beta - D^{-1}\upsilon.$$

The matrix X is the design matrix with rows X_i^T, I is an $(n \times n)$ identity matrix,

$$\boldsymbol{\upsilon} = (\dot{\ell}_1, \dots, \dot{\ell}_n), \quad (11)$$

$$\mathbf{D} = \mathbf{diag}(\ddot{\ell}_1, \dots, \ddot{\ell}_n), \quad (12)$$

and S^P is a smoother matrix with elements

$$(\mathbf{S^P})_{ij} = \frac{\ddot{\ell}_{ij} \mathcal{K}_H (W_i - W_j)}{\sum_{i=1}^{n} \ddot{\ell}_{ij} \mathcal{K}_H (W_i - W_j)} \quad (13)$$

Third: The updating step for $m_j = m_\beta(w_j)$

The function $m_j = m_\beta(w_j)$ is updated by

$$\hat{m}_j^{k+1} = \hat{m}_j^k \frac{\sum_{i=1}^n \ell'_{ij} \mathcal{K}_H (W_i - W_j)}{\sum_{i=1}^n \ell''_{ij} \mathcal{K}_H (W_i - W_j)},$$

where k=0, l, 2,...is the number of iteration.

It is noted that the function $\ell''_{ij}(.)$ can be replaced by its expectation with respect to Y to obtain a Fisher scoring type algorithm (Severini and Staniswalis, 1994).

The previous procedure can be summarized as follows.

Updating step for β

$$\beta^{new} = (\tilde{X}^T D \tilde{X})^{-1} \tilde{X}^T D \tilde{Z}.$$

Updating step for m_j

$$m_j^{new} = m_j \frac{\sum_{i=1}^n \ell'_{ij} \mathcal{K}_H (W_i - W_j)}{\sum_{i=1}^n \ell''_{ij} \mathcal{K}_H (W_i - W_j)}.$$

Notes on the procedure:

1. The variable $\tilde{Z}$, which is defined here, is a set of adjusted dependent variable.
2. The parameter β is updated by a parametric method with a nonparametrically modified design matrix $\tilde{X}$.
3. The function ℓ''_i can be replaced by its expectation, with respect to y, to obtain a Fisher scoring type procedure.
4. The updating step for m_j is of quite complex structure and can be simplified in some models for identity and exponential link functions G.

2.2. Generalized Speckman Method

Generalized Speckman estimation method back to Speckman (1988). In the case of identity link function G and normally distributed Y, the generalized Speckman and profile likelihood methods coincides with no updating steps are needed for the estimation of both $\underline{\beta}$ and *m*. This method can be summarized in the case of identity link (PLM) as follows:

(1) Estimate *p* by: $\widehat{\beta} = (\widetilde{X}^T\widetilde{X})^{-1}\widetilde{X}^T\widetilde{Y}$

(2) Estimate *m* by: $\widehat{m} = S(Y - X\widehat{\beta})$, where

$\widetilde{X} = (1 - S)X$, $\widetilde{Y} = (1 - S)Y$, and a smoother matrix S is defined by its elements as $(S)_{ij} = \frac{\mathcal{K}_H(W_i - W_j)}{\sum_{i=1}^{n}\mathcal{K}_H(W_i - W_j)}$.

This matrix is a simpler form of a smoother matrix, and differs from the one used in Eq. (13) where the matrix **S**yields $\ell'_i = -(Y_i - X_i^T\beta - m_j)/\sigma^2$ and $\ell''_i = -1/\sigma^2$, in the case of normally distributed Y.

For the GPLM, the Speckman estimator is combined with the IWLS method used in the estimation of GLM. As it is shown in IWLS each iteration step of GLM was obtained by WLS regression on the adjusted dependent variable. The same procedure will be used in the GPLM by replacing IWLS with a weighted partial linear fit on the adjusted dependent variable given by

$$Z = X\beta + m - D^{-1}\upsilon,$$

where υ and D are defined as in Equations (11) and (12) respectively. The generalized Speckman algorithm for the GPLM can be summarized as:

First: Initial values:

The initial values used in this method are the same as in the previous profile likelihood algorithm.

Second: Updating step for $\underline{\beta}$

$$\beta^{new} = (\widetilde{X}^T D\widetilde{X})^{-1}\widetilde{X}^T D\widetilde{Z}.$$

Third: Updating step for m

$$m^{new} = S(Z - X\beta),$$

Where

$$\widetilde{U} = (1 - S)X,$$

$$\widetilde{Z} = \widetilde{X}\beta - D^{-1}\upsilon \ ,$$

The smoother matrix is used with elements:

$$(\mathbf{S})_{ij} = \frac{\ell''_i \, \mathcal{K}_{\mathbf{H}}(\mathbf{w}_i - \mathbf{w}_j)}{\sum_{i=1}^{n} \ell''_i \, \mathcal{K}_{\mathbf{H}}(\mathbf{w}_i - \mathbf{w}_j)} \tag{14}$$

There is a difference between the smoother matrix in Equation (14) and that in Equation (13). In Equation (13) we use ℓ''_i instead of ℓ''_{ij} that is used in Equation (14).

2.3. Back-fitting Method

Hastie and Tibishirani (1990) suggested the back-fitting method as an iterative algorithm to fit an additive model. The idea of this method is to regress the additive components separately on the partial residuals.

The back-fitting method will be presented in the case of identity link G (PLM) and non-monotone G (GPLM) as follows:

Back-fitting algorithm for the GPLM

Back-fitting for the GPLM is an extension to that of PLM. The iterations in this method coincide with that in the Speckman method. This method differs from the Speckman method only in the parametric part. The back-fitting algorithm for the GPLM can be summarized as follows

First: Initial values:

This method often use

$$\boldsymbol{\beta} = \mathbf{0}, \text{ and } \mathbf{m}(\mathbf{w}_j) = \mathbf{G}^{-1}(\overline{\mathbf{Y}})$$

Second: Updating step for $\boldsymbol{\beta}$

$$\boldsymbol{\beta}^{\mathbf{new}} = \left(\widetilde{\mathbf{X}}^{\mathbf{T}}\mathbf{D}\widetilde{\mathbf{X}}\right)^{-1}\widetilde{\mathbf{X}}^{\mathbf{T}}\mathbf{D}\widetilde{\mathbf{Z}}.$$

Third: Updating step for m

$$\mathbf{m}^{\mathbf{new}} = \mathbf{S}(\mathbf{Z} - \mathbf{X}\boldsymbol{\beta}),$$

where the matrices D and S, the vector v, $\widetilde{X}$ and $\widetilde{Z}$ are defined as in the Speckman method for the GPLM (See Muller, 2001).

In practice, some of the predictor variables are correlated. Therefore, Hastie and Tibshirani (1990) proposed a modified back-fitting method which first search for a parametric solution and only fit the remaining parts non-parametrically.

2.4. Some Statistical Properties of the GPLM Estimators.

(1) Statistical properties of the parametric component β

Under some regularity conditions the estimator β has the following properties:

1. 1. $\sqrt{n}$-consistent estimator for β.
2. Asymptotically normal.
3. Its limiting covariance has a consistent estimator.
4. Asymptotically efficient; has asymptotically minimum variance (Severini and Staniswalis (1994).

(2) Statistical properties of the non-parametric component m:

The non-parametric function m can be estimated (in the univariate case) with the usual univariate rate of convergence. Severini and Staniswalis (1994) showed that the estimator $\widehat{m}_{\beta}(.)$ is consistent in supremum norm. They showed that the parametric and non-parametric estimators have the following asymptotic properties:

$$\widehat{\boldsymbol{\beta}} = \boldsymbol{\beta}_o + O_p\left(n^{(\frac{-1}{2})}\right),$$

$$\left\|\widehat{\mathbf{m}}_{\underline{\boldsymbol{\beta}}} - \mathbf{m}_o\right\| = O_p\left(n^{(\frac{-1}{4})}\right), \text{ as } n \to \infty,$$

$$\widehat{\boldsymbol{\sigma}}^2 = \widehat{\boldsymbol{\sigma}}_o^2 + O_p(1),$$

where $\underline{\beta_o}, m_o, \widehat{\sigma}_o^2$ are the true parameter values so that

$$\mathbf{var}(\mathbf{Y}|\mathbf{X}, \mathbf{W}) = \widehat{\boldsymbol{\sigma}}_o^2 \ \mathbf{var}\left[G\left(\mathbf{X}\boldsymbol{\beta}_o + \mathbf{m}_o(\mathbf{W})\right)\right].$$

3. BAYESIAN ESTIMATION AND INFERENCE FOR THE GPLM

Bayesian inference derives the posterior distribution as a consequence of two antecedents; a prior probability and a likelihood function derived from a probability model for the data to be observed. In Bayesian inference the posterior probability can be obtained according to Bayes theorem as follows

$$P(\boldsymbol{\beta}|\mathbf{y}) \propto \pi(\boldsymbol{\beta})\mathrm{L}(\mathbf{y}|\boldsymbol{\beta}), \tag{15}$$

where $P(\boldsymbol{\beta}|\mathbf{y})$ is the posterior distribution, $\pi(\boldsymbol{\beta})$ is the prior distribution **and** $\mathrm{L}(\mathbf{y}|\boldsymbol{\beta})$ is the likelihood function.

3.1. A proposed Algorithm for Estimating the GPLM Parameters

1. Obtain the probability distribution of response variable, $\underline{\mathrm{Y}}$.
2. Obtain the likelihood function of the probability distribution of response variable $\underline{\mathrm{Y}}$.
3. Choose a suitable prior distribution of $\boldsymbol{\beta}$.
4. Use Eq. (15) to obtain the posterior distribution.
5. Obtain the Bayesian estimator under the square error loss function.
6. Replace the initial value of $\boldsymbol{\beta}$ by the Bayesian estimator.
7. Use the profile likelihood method, generalized Speckman method and Back-fitting method with the new initial value of $\boldsymbol{\beta}$ to estimate the GPLM parameters.
8.

3.2. Bayesian approach for Estimating the GPLM Using Bayesian Estimator

3.2.1. σ^2 is assumed known

Case 1

Consider the GPLM in Eq. (1) and suppose that $\underline{Y}$ belongs to the multivariate normal distribution with pdf

$$f(y_i|X_i,\beta,\sigma^2)=\frac{1}{\sigma\sqrt{2\pi}}\exp\frac{-(y_i-\beta^T x_i)^2}{2\sigma^2},$$

Then the likelihood function of the variable $\underline{Y}$ can be written as

$$\prod_{i=1}^{n} f(y_i|X_i,\beta,\sigma^2)\propto \exp\left[\frac{-1}{2\sigma^2}.\,SSR(\beta)\right], \quad (16)$$

where

$$SSR(\beta)=(-2\beta^T X^T y+\beta^T X^T X\beta).$$

Let

$\underline{\beta}$ ~ Multivariate Nomal Distribution ($\underline{\mu}$, Σ)

Then

$$f(\beta)=(2\pi)^{-P/2}|\Sigma|^{-1/2}\exp\left(\frac{-1}{2}\left(\beta-\underline{\mu}\right)^T\Sigma^{-1}\left(\beta-\underline{\mu}\right)\right), \quad (17)$$

Combining (16) and (17) and using (15), we obtain the following posterior distribution (using some algebraic steps)

$$P(\beta|Y,X,\sigma^2)\propto\exp\left[-\frac{1}{2}\left(\Sigma^{-1}+\frac{X^TX}{\sigma^2}\right)^{-1}\left(\beta-\left(\Sigma^{-1}+\frac{X^TX}{\sigma^2}\right)^{-1}\left(\Sigma^{-1}\underline{\mu}+\frac{X^TY}{\sigma^2}\right)\right)\right],$$

which is multivariate normal distribution with

$$E(\beta|Y,X,\sigma^2)=\left(\Sigma^{-1}+X^TX/\sigma^2\right)^{-1}\left(\Sigma^{-1}\underline{\mu}+X^TY/\sigma^2\right),$$

$$Var(\beta|Y,X,\sigma^2)=\left(\Sigma^{-1}+X^TX/\sigma^2\right)^{-1}.$$

Then, the Bayesian estimator under the square error loss function

$$\hat{\beta}_{(bayes)}=\left(\Sigma^{-1}+X^TX/\sigma^2\right)^{-1}\left(\Sigma^{-1}\underline{\mu}+X^TY/\sigma^2\right). \quad (18)$$

Case 2

Consider the GPLM in Eq. (1) and suppose that the pdf of $\underline{Y}$ as follows

$$(2\pi)^{\frac{-p}{2}}|\Sigma|^{-1/2}\exp\left\{-(X_i-\beta)^T\Sigma^{-1}(X_i-\beta)/2\right\}.$$

Then the likelihood function of the probability distribution of $\underline{Y}$ is

$$((2\pi)^{\frac{-np}{2}}|\Sigma|^{-n/2} \exp\left\{-\frac{1}{2}\sum_{i=1}^{n}(X_i-\beta)^T\Sigma^{-1}(X_i-\beta)\right\}$$

$$= \exp\left\{-\frac{1}{2}\beta^T n\Sigma^{-1}\beta + \beta^T n\Sigma^{-1}\left(\frac{1}{n}\sum_{i=1}^{n}X_{i,1}, \sum_{i=1}^{n}X_{i,2} \ldots, \sum_{i=1}^{n}X_{i,p}\right)^T\right\}.$$

Let

$$A_1 = n\Sigma^{-1},$$

$$b_1 = n\Sigma^{-1}\overline{X},$$

$$\overline{X} = \left(\frac{1}{n}\sum_{i=1}^{n}X_{i,1}, \sum_{i=1}^{n}X_{i,2} \ldots, \sum_{i=1}^{n}X_{i,p}\right)^T.$$

Then, we can rewrite the likelihood function of as follows

$$\propto \exp\left\{-\frac{1}{2}\beta^T A_1\beta + \beta^T b_1\right\}, \tag{19}$$

Let $\underline{\beta}\sim$ multivariate normal distribution (μ_0, Λ_0),

Then

$$\begin{aligned}
\pi(\beta) &= (2\pi)^{\frac{-p}{2}}|\Lambda_0|^{-1/2}\exp\left\{-\frac{1}{2}(\beta-\mu_0)^T\Lambda_0^{-1}(\beta-\mu_0)\right\} \\
&= (2\pi)^{\frac{-p}{2}}|\Lambda_0|^{-1/2}\exp\left\{-\frac{1}{2}\beta^T\Lambda_0^{-1}\underline{\beta} + \beta^T\Lambda_0^{-1}\mu_0 - \frac{1}{2}\mu_0^T\Lambda_0^{-1}\mu_0\right\} \\
&\propto \exp\left\{-\frac{1}{2}\beta^T\Lambda_0^{-1}\underline{\beta} + \underline{\beta}^T\Lambda_0^{-1}\mu_0\right\} \\
&\propto \exp\left\{-\frac{1}{2}\beta^T A_1\beta + \beta^T b_1\right\},
\end{aligned} \tag{20}$$

where

$$A_1 = \Lambda_0^{-1},$$
$$b_1 = \Lambda_0^{-1}\mu_0,$$
$$A_0 = \Lambda_0^{-1},$$
$$b_0 = \Lambda_0^{-1}\mu_0.$$

Combining (19) and (20) and using (15), we obtain the following posterior distribution (using some algebraic steps)

$$\propto \exp\left\{-\frac{1}{2}\beta^T A_1\beta + \beta^T b_1\right\} \times \exp\left\{-\frac{1}{2}\beta^T A_1\beta + \beta^T b_1\right\} = \exp\left\{-\frac{1}{2}\beta^T A_n\beta + \beta^T b_n\right\},$$

where

$$\mathbf{A_n} = \mathbf{A_0} + \mathbf{A_1} = \boldsymbol{\Lambda_0}^{-1} + \mathbf{n}\,\boldsymbol{\Sigma}^{-1},$$

$$\mathbf{b_n} = \mathbf{b_0} + \mathbf{b_1} = \boldsymbol{\Lambda_0}^{-1}\boldsymbol{\mu_0} + \mathbf{n}\boldsymbol{\Sigma}^{-1}\overline{\mathbf{X}},$$

which is multivariate normal distribution with

$$\mathbf{E}\left(\boldsymbol{\beta}\middle|\mathbf{Y},\mathbf{X},\boldsymbol{\Sigma}\right) = \mathbf{A_n}^{-1}\,\mathbf{b_n} = \left(\boldsymbol{\Lambda_0}^{-1} + \mathbf{n}\,\boldsymbol{\Sigma}^{-1}\right)^{-1}\left(\boldsymbol{\Lambda_0}^{-1}\boldsymbol{\mu_0} + \mathbf{n}\boldsymbol{\Sigma}^{-1}\overline{\mathbf{X}}\right)$$

and

$$\mathbf{COV}\left(\boldsymbol{\beta}\middle|\mathbf{Y},\mathbf{X},\boldsymbol{\Sigma}\right) = \boldsymbol{\Lambda_n} = \left(\boldsymbol{\Lambda_0}^{-1} + \mathbf{n}\,\boldsymbol{\Sigma}^{-1}\right)^{-1}.$$

Then, the Bayesian estimator under the square error loss function **is**

$$\widehat{\boldsymbol{\beta}}_{(\mathbf{bayes})} = \left(\boldsymbol{\Lambda_0}^{-1} + \mathbf{n}\,\Sigma^{-1}\right)^{-1}\left(\boldsymbol{\Lambda_0}^{-1}\boldsymbol{\mu_0} + \mathbf{n}\Sigma^{-1}\overline{\mathbf{X}}\right) \qquad (21)$$

3.2.2. $\underline{\beta}$ and σ^2 is assumed unknown

Case 1

Consider the GPLM in Eq. (1) and suppose that the marginal posterior distribution for σ^2 is proportional with $f(\sigma^2)$ and $f(Y|X,\sigma^2)$ so that

$$\boldsymbol{\pi}(\boldsymbol{\sigma}^2\,|\mathbf{Y},\mathbf{X}) \propto \boldsymbol{\pi}(\boldsymbol{\sigma}^2).\boldsymbol{\pi}(\mathbf{Y}|\mathbf{X},\boldsymbol{\sigma}^2). \qquad (22)$$

Hence

$$\boldsymbol{\pi}(\mathbf{Y}|\mathbf{X},\boldsymbol{\sigma}^2) = \int \boldsymbol{\pi}(\mathbf{Y}|\mathbf{X},\boldsymbol{\beta},\boldsymbol{\sigma}^2)\boldsymbol{\pi}(\boldsymbol{\beta}|\mathbf{X},\boldsymbol{\sigma}^2)\;\; \mathbf{d}\,\boldsymbol{\beta},$$

where

$$\boldsymbol{\pi}(\mathbf{Y}|\mathbf{X},\boldsymbol{\beta},\boldsymbol{\sigma}^2) = (2\boldsymbol{\pi}\boldsymbol{\sigma}^2)^{-n/2}\exp\left[\frac{-1}{2\sigma^2}(\mathbf{y}-\mathbf{X}\boldsymbol{\beta})^{\mathrm{T}}(\mathbf{y}-\mathbf{X}\boldsymbol{\beta})\right]$$

$$\boldsymbol{\pi}(\boldsymbol{\beta}|\mathbf{X},\boldsymbol{\sigma}^2) = |2\boldsymbol{\pi}\mathbf{g}\boldsymbol{\sigma}^2(\mathbf{X}^{\mathrm{T}}\mathbf{X})^{-1}|^{-1}\exp\left[\frac{-1}{2g\sigma^2}\boldsymbol{\beta}^{\mathrm{T}}\mathbf{X}^{\mathrm{T}}\mathbf{X}\boldsymbol{\beta}\right].$$

Subsequently

$$\pi(Y|X,\sigma^2) = \int \left\{(2\pi\sigma^2)^{-\frac{n}{2}} \exp\left[\frac{-1}{2\sigma^2}(y - X\beta)^T(y - X\beta)\right]\right\}\left\{|2\pi g\sigma^2 (X^TX)^{-1}|^{-1} \exp\left[\frac{-1}{2g\sigma^2}\beta^T X^T X\beta\right]\right\} d\beta$$

$$\pi(Y|X,\sigma^2) = \int \left\{(2\pi\sigma^2)^{-\frac{n}{2}}\right\}\{|2\pi g\sigma^2 (X^TX)^{-1}|^{-1}\} \exp\left\{\frac{-1}{2\sigma^2}[(y - X\beta)^T(y - X\beta)\beta^T X^T X\beta/g]\right\} d\beta$$

$$\pi(Y|X,\sigma^2) = \int \left\{(2\pi\sigma^2)^{-\frac{n}{2}}\right\}\{|2\pi g\sigma^2 (X^TX)^{-1}|^{-1}\} \exp\left\{\frac{-1}{2\sigma^2}\left[y^Ty - 2y^TX\beta + \beta^T X^T X\beta(1 + \frac{1}{g})\right]\right\} d\beta$$

$$\pi(Y|X,\sigma^2) = \int \left\{(2\pi\sigma^2)^{-\frac{n}{2}}\right\}\{|2\pi g\sigma^2 (X^TX)^{-1}|^{-1}\} \exp\left\{\frac{-1}{2\sigma^2}y^Ty - \frac{1}{2}(\beta - M)^T V^{-1}(\beta - M) + \frac{1}{2}M^TV^{-1}M\right\} d\beta,$$

where

$$M = \frac{g}{1+g}\ (X^TX)^{-1}\ X^Ty,\quad V = \frac{g}{1+g}\sigma^2(X^TX)^{-1},\ g = \frac{K}{\sigma^2}.$$

K is any positive number, then we can rewrite the last expression as follows

$$\pi(Y|X,\sigma^2) = \left\{(2\pi)^{-\frac{n}{2}}\right\} \times \left\{(1+g)^{\frac{-p}{2}}\right\} \times \left\{(\sigma^2)^{-\frac{n}{2}}\right\} \times \left\{\exp\left[\frac{-1}{2\sigma^2}(SSR_g)\right]\right\}. \tag{23}$$

Let

$$\gamma = \frac{1}{\sigma^2} \sim \text{Gamma}\left(\frac{\upsilon_0}{2}, \frac{\upsilon_0\sigma_0^2}{2}\right)$$

Then

$$\pi(\gamma) \propto \left\{\gamma^{\frac{\upsilon_0}{2}-1} \exp\left[-\gamma\frac{\upsilon_0\sigma_0^2}{2}\right]\right\}, \tag{24}$$

Substituting form (24) and (25) into (23) we get

$$\pi(\gamma|Y,X) \propto \left\{\gamma^{\frac{\upsilon_0}{2}-1} \exp\left[-\gamma\frac{\upsilon_0\sigma_0^2}{2}\right]\right\}\left\{\gamma^{\frac{n}{2}} \exp\left[-\gamma\frac{(SSR_g)}{2}\right]\right\}.$$

Therefore

$$\pi(\gamma|Y,X) \propto \left\{\gamma^{\frac{\upsilon_0+n}{2}-1} \exp - \frac{\gamma[\upsilon_0\sigma_0^2 + (SSR_g)]}{2}\right\}.$$

Then, the posterior distribution

$$\pi(\gamma|Y,X) \sim \text{Gamma}\left\{\frac{\upsilon_0 + n}{2}, \frac{\upsilon_0\sigma_0^2 + (SSR_g)}{2}\right\}$$

and the marginal posterior distribution for σ^2 is

$$\pi(\sigma^2|Y,X) \sim \textbf{Inverse Gamma}\left\{\frac{\upsilon_0+n}{2}, \frac{\upsilon_0\sigma_0^2+(SSR_g)}{2}\right\} \tag{25}$$

Case 2

Consider the GPLM in Eq. (1) and suppose that

$$f(y_i|X,\beta,\sigma^2) = \frac{1}{\sigma\sqrt{2\pi}} \exp \frac{-\left(y_i-\underline{\beta}^T x_i\right)^2}{2\sigma^2} \; .$$

Then the likelihood function of the pdf of response variable $\underline{Y}$

$$\prod_{i=1}^{n} f(y_i|X_i,\beta,\sigma^2) \propto \exp\left[\frac{-1}{2\sigma^2} . \mathrm{SSR}\left(\underline{\beta}\right)\right], \tag{26}$$

where

$$\mathrm{SSR}(\beta) = (-2\beta^T X^T y + \beta^T X^T X\beta).$$

Let

$$\underline{\beta} \sim \text{Multivariate Nomal Distribution } (\mu_\beta, \sigma^2 V_\beta) \tag{27}$$

and

$$\sigma^2 \sim \text{Inverse Gamma } (a, b), \tag{28}$$

Then from (27) and (28) we get

$$\pi\ (\beta,\sigma^2) = N(\mu_\beta, \sigma^2 V_\beta) \times IG(a,b) = \mathrm{MNIG}\ (\mu_\beta, V_\beta, a, b),$$

$$\pi\ (\beta,\sigma^2) = \frac{b^a}{(2\pi)^{\frac{p}{2}} |V_\beta|^{\frac{1}{2}} \Gamma(a)} \left(\frac{1}{\sigma^2}\right)^{\frac{2+p}{2}+1} \times \exp\left[-\frac{1}{\sigma^2}\left\{b + \frac{1}{2}(\beta-\mu_\beta)^T V_\beta^{-1}(\beta-\mu_\beta)\right\}\right]$$

$$\pi\ (\beta,\sigma^2) \propto \left(\frac{1}{\sigma^2}\right)^{\frac{2+p}{2}+1} \times \exp\left[-\frac{1}{\sigma^2}\left\{b + \frac{1}{2}(\beta-\mu_\beta)^T V_\beta^{-1}(\beta-\mu_\beta)\right\}\right],\ a, b, \sigma^2 > 0 \tag{29}$$

Combining (26) and (29), using (15) and after some algebraic steps, we obtain the following joint posterior distribution

$$\pi(\beta,\sigma^2|Y) \propto \left(\frac{1}{\sigma^2}\right)^{\frac{2+p}{2}+1} \exp\left[-\frac{1}{\sigma^2}\left\{b^* + \frac{1}{2}(\beta-\mu^*)^T V^{*-1}(\beta-\mu^*)\right\}\right], \tag{30}$$

where

$$\mu^* = (V_\beta^{-1} + X^TX)^{-1} (V_\beta^{-1} \mu_\beta + X^TY),$$
$$V^* = (V^{-1} + X^TX)^{-1},$$
$$a^* = a + n/2,$$
$$b^* = b + \frac{1}{2}[\mu_\beta{}^T V_\beta^{-1} + Y^TY + \mu^{*T} V^{*-1} \mu^*].$$

Then

$$\pi(\beta, \sigma^2 | Y) \sim \text{NIG}(\mu^*, V^*, a^*, b^*),$$

which is normal inverse gamma distribution.

From (30), the marginal posterior distribution for σ^2

$$\pi(\beta|Y) = \int \pi(\beta, \sigma^2|Y)\, d\beta = \int_\beta \text{NIG}(\mu^*, V^*, a^*, b^*) d\beta$$

$$\pi(\sigma^2|Y) = \frac{b^{*a^*}}{\Gamma(a^*)} \left(\frac{1}{\sigma^2}\right)^{a^*+1} \exp\left[-\frac{b^*}{\sigma^2}\right], \quad \pi(\sigma^2|Y) \sim \text{NIG}(a^*, b^*). \tag{31}$$

From (30), the marginal posterior distribution for β

$$\pi(\beta|Y) = \int \pi(\beta, \sigma^2|Y)\, d\sigma^2 = \int_{\sigma^2} \text{NIG}(\mu^*, V^*, a^*, b^*) d\sigma^2$$

$$\propto \int \left(\frac{1}{\sigma^2}\right)^{\frac{a+n}{2}+1} \times \exp\left[-\frac{1}{\sigma^2}\left\{b^* + \frac{1}{2}(\beta - \mu^*)^T V^{*-1} (\beta - \mu^*)\right\}\right] \, d\sigma^2$$

$$\propto \left\{1 + \frac{(\beta-\mu^*)^T V^{*-1} (\beta-\mu^*)}{2b^*}\right\}^{-\left(\frac{2a^*+p}{2}\right)}, \tag{32}$$

which is multivariate t-distribution with the following pdf

$$\frac{\Gamma\left(\frac{\upsilon^*+p}{2}\right)}{\Gamma\left(\frac{\upsilon^*}{2}\right) \pi^{\frac{p}{2}} \left|\upsilon^* \Sigma^*\right|^{\frac{1}{2}}} \left\{1 + \frac{(\beta - \mu^*)^T \Sigma^{*-1} (\beta - \mu^*)}{\upsilon^*}\right\}^{-\left(\frac{\upsilon^*+p}{2}\right)}$$

with expected value $= 0$, where

$$\upsilon^* = 2a^*$$

$$\Sigma^* = \left(\frac{b^*}{a^*}\right)^{V^*}.$$

Then, the Bayesian estimator under the square error loss function

$$\hat{\beta}_{(bayes)} = 0. \tag{33}$$

3.2.3. $\underline{\beta}$ is assumed known

Case 1

Consider the GPLM in Eq. (1) with likelihood function of the pdf of response variable $\underline{Y}$

$$\prod_{i=1}^{n} f(Y|X,\beta,\gamma) \propto \left\{\gamma^{\frac{n}{2}} \exp\left[-\gamma \frac{SSR(\beta)}{2}\right]\right\}. \tag{34}$$

Let

$$\gamma = \frac{1}{\sigma^2} \sim \text{Gamma}\left(\frac{\upsilon_0}{2}, \frac{\upsilon_0 \sigma_0^2}{2}\right).$$

Then

$$\pi(\gamma) \propto \left\{\gamma^{\frac{\upsilon_0}{2}-1} \exp\left[-\gamma \frac{\upsilon_0 \sigma_0^2}{2}\right]\right\}, \tag{35}$$

Combining (34) and (35) Using (15) and after some algebra, we obtain the following posterior distribution

$$\pi(\gamma|Y,X,\beta) \propto \left\{\gamma^{\frac{\upsilon_0+n}{2}-1} \exp -\frac{\gamma\left[\upsilon_0\sigma_0^2 + SSR(\beta)\right]}{2}\right\},$$

$$(\sigma^2|Y,X,\beta) \sim \text{Inverse Gamma}\left\{\frac{\upsilon_0+n}{2}, \frac{\upsilon_0\sigma_0^2+SSR(\beta)}{2}\right\}. \tag{36}$$

Case 2

Consider the GPLM in Eq. (1) with likelihood function of the pdf of response variable $\underline{Y}$

$$(2\pi)^{\frac{-np}{2}} |\Sigma|^{-n/2} \exp\left\{-\frac{1}{2}\sum_{i=1}^{n}(X_i - \beta)^T \Sigma^{-1}(X_i - \beta)\right\}, \tag{37}$$

where

$$\sum_{i=1}^{n}(X_i - \beta)^T \Sigma^{-1} (X_i - \beta) = tr\left(S_\beta \Sigma^{-1}\right),$$

$$S_\beta = \sum_{i=1}^{n}(X_i - \beta)^T (X_i - \beta).$$

Let

$$\pi(\Sigma) \sim \text{Inverse Wishert}\{\upsilon_0, S_0^{-1}\}, \tag{38}$$

Combining (37) and (38) Using (15) and after some algebra, we obtain the following posterior distribution

$$\propto \left[|\Sigma|^{-\left(\frac{\upsilon_0+P+1}{2}\right)} \times \exp\left\{-\mathrm{tr}\left(\frac{S_0\Sigma^{-1}}{2}\right)\right\}\right]\left[|\Sigma|^{-n/2}\exp\left\{-\frac{1}{2}\mathrm{tr}\left(S_\beta\Sigma^{-1}\right)\right\}\right]$$
$$=|\Sigma|^{-\frac{(\upsilon_0+P+n+1)}{2}}\exp\{-\mathrm{tr}[(S_0+S_\beta)\Sigma^{-1}]/2\},$$
(39)

which is inverse Wishert distribution where

$$\left(\Sigma\middle|Y,U,\beta\right)\sim\text{Inverse Wishert}\left\{\upsilon_0+n,\left(S_0+S_\beta\right)^{-1}\right\}$$

with expected value

$$\frac{1}{\upsilon_0-P+n+1}\left(S_0+S_\beta\right).$$

Then, the Bayesian estimator under the square error loss function

$$\hat{\Sigma}_{(bayes)}=\frac{\upsilon_0-P-1}{\upsilon_0-P+n+1}\frac{1}{\upsilon_0-P-1}S_0+\frac{n}{\upsilon_0-P+n+1}\frac{1}{n}S_\beta$$

The previous results are summarized in Table (1).

Table (1). The Posterior Distribution Functions

Probability distribution of Y	Unknown parameter	Conjugate prior distribution	Posterior distribution	
Multivariate Normal Distribution	β	Multivariate Normal Distribution	Multivariate Normal Distribution	
Multivariate Normal Distribution	$(\sigma^2)^{-1}$	(GD)	(IGD)	
Multivariate Normal Distribution	β	(MND)	Joint distribution	(NIG)
	σ^2	(IG)	Marginal distribution for β	(MStD)
Multivariate Normal Distribution	σ^2	(IWD)	(IWD)	

(NIG)= Normal Inverse Gamma, (IGD) = Inverse Gamma, (MStD) = Multivariate Student t Distribution, (IWD) = Inverse Wishert Distribution, (GD) = Gamma Distribution.

4. SIMULATION STUDY

The aim of this simulation study is twofold. First, is to evaluate the performance of the proposed Bayesian estimators. Second, is to compare the proposed technique with classical approaches. The simulation study is based on 500 Monte Carlo replications. Different sample sizes have been used ranging from small, moderate, to large sizes. In specific sample sizes are fixed at n=50, n=100, n=200, n=500, n=100, and n=2000. Also, different bandwidth parameters are used, namely, h=2, h=1, h=0.5, and h=0.2.

The estimation results and statistical analysis are obtained using statistical Package XploRe, 4.8 (XploRe, 2000).

The methods are compared according to the following criteria.

First, the Average Mean of Squared Errors for $(\hat{\beta})$, AMSE $(\hat{\beta})$, where

$$\mathbf{AMSE}\ (\hat{\boldsymbol{\beta}}) = \mathbf{n}^{-1} \sum_{\mathbf{i=1}}^{\mathbf{n}} \mathbf{E}\{\boldsymbol{\beta}_{\mathbf{i}} - \hat{\boldsymbol{\beta}}_{\mathbf{i}}\}^{\mathbf{2}}.$$

The second is the Average Mean of Squared Errors for $(\hat{m})$, AMSE $(\hat{m})$, where

$$\mathbf{AMSE}\ (\hat{\mathbf{m}}) = \mathbf{n}^{-1} \sum_{\mathbf{i=1}}^{\mathbf{n}} \mathbf{E}\{\mathbf{m}_{\mathbf{i}} - \hat{\mathbf{m}}_{\mathbf{i}}\}^{\mathbf{2}}.$$

The third is the deviance where,

Deviance = -2 Log Likelihood.

The results are shown in the following Tables (2) – (7).

Table (2). Simulation Results for n = 50

Criterion		n = 50					
		Classical Estimation			Bayesian Estimation		
Bandwidth and model		AMSE $(\hat{\beta})$	AMSE $(\hat{m})$	Mean Of Deviance	AMSE $(\hat{\beta})$	AMSE $(\hat{m})$	Mean Of Deviance
h=2	P- Likelihood	0.8271	0.8815	33.38	0.8215	0.8469	32.82
	G- Speckman	0.9473	0.7543	34.39	0.8849	0.7371	32.73
	Backftting	0.7874	0.7332	33.32	0.71435	0.7276	30.37
h=1	P- Likelihood	0.8291	0.7803	45.12	0.8053	0.7049	43.64
	G- Speckman	0.9464	0.7161	45.39	0.8128	0.7088	43.11
	Backftting	0.7859	0.7039	43.45	0.7544	0. 6631	41.66
h=0.5	P- Likelihood	0.7712	0.6966	59.58	0.6649	0.6194	56.11
	G- Speckman	0.8981	0.6325	58.51	0.7146	0.6267	57.35
	Backftting	0.7643	0.6120	57.81	0.6616	0.6120	56.01
h=0.2	P- Likelihood	0.7631	0.4929	68.35	0. 6132	0.4052	64.98
	G- Speckman	0.7854	0.4341	68.59	0.6245	0. 4181	66.11
	Backftting	0.7562	0.4127	66.61	0.6087	0.3758	64.77

Table (3). Simulation Results for n = 100

Criterion / Bandwidth and model		n = 100					
		Classical Estimation			Bayesian Estimation		
		AMSE ($\hat{\beta}$)	AMSE ($\hat{m}$)	Mean Of Deviance	AMSE ($\hat{\beta}$)	AMSE ($\hat{m}$)	Mean Of Deviance
h=2	P- Likelihood	0.8043	0.8738	49.94	0.7313	0. 7401	43.76
	G- Speckman	0.9144	0.8399	46.87	0.7377	0.7434	41.54
	Backfitting	0.7778	0.8232	44.35	0.7061	0.7322	40.12
h=1	P- Likelihood	0.8011	0.7268	55.89	0.6864	0. 6746	54.41
	G- Speckman	0.9131	0.6351	57.69	0.6949	0.6459	54.36
	Backfitting	0.7749	0.6338	59.57	0.6533	0.5676	53.51
h=0.5	P- Likelihood	0.7810	0.5552	71.33	0.6155	0.4577	66.57
	G- Speckman	0.8612	0.5569	70.89	0. 6147	0.4534	66.63
	Backfitting	0.7718	0.5435	70.85	0.6014	0. 4293	64.22
h=0.2	P- Likelihood	0.7689	0.3346	78.66	0.5661	0.2489	75.91
	G- Speckman	0.7768	0.3473	79.75	0.5642	0.2433	75.11
	Backfitting	0.7661	0.3308	80.21	0.5457	0.2176	74.66

Table (4). Simulation Results for n = 200

Criterion / Bandwidth and model		n = 200					
		Classical Estimation			Bayesian Estimation		
		AMSE ($\hat{\beta}$)	AMSE ($\hat{m}$)	Mean Of Deviance	AMSE ($\hat{\beta}$)	AMSE ($\hat{m}$)	Mean Of Deviance
h=2	P- Likelihood	0.7523	0.8610	52.99	0.6553	0.6540	45.91
	G- Speckman	0.8643	0.7732	52.93	0.6611	0.6581	45.79
	Backfitting	0.7545	0.7954	52.76	0.6549	0.6521	45.76
h=1	P- Likelihood	0.7544	0.6961	70.81	0.5256	0.5133	66.43
	G- Speckman	0.8667	0.6016	73.01	0.5298	0.5216	67.51
	Backfitting	0.7528	0.6265	72.53	0.5332	0.5278	67.79
h=0.5	P- Likelihood	0.7465	0.4368	84.36	0.4989	0. 3276	81.54
	G- Speckman	0.8389	0.5063	81.44	0.4959	0.3546	80.28
	Backfitting	0.7345	0.4829	80.75	0.4945	0.3258	80.21
h=0.2	P- Likelihood	0.7034	0.2342	90.21	0.4503	0.1839	88.11
	G- Speckman	0.8036	0.2973	91.33	0.4621	0.2059	89. 38
	Backfitting	0.7043	0.2896	90.99	0.4635	0.2046	89. 86

Table (5). Simulation Results for n = 500

Criterion / Bandwidth and model		n = 500					
		Classical Estimation			Bayesian Estimation		
		AMSE ($\hat{\beta}$)	AMSE ($\hat{m}$)	Mean Of Deviance	AMSE ($\hat{\beta}$)	AMSE ($\hat{m}$)	Mean Of Deviance
h=2	P- Likelihood	0.5466	0.7067	163.92	0. 4345	0.5822	160.34
	G- Speckman	0.6578	0.7149	164.88	0.4371	0.5876	162.37
	Backftting	0.5756	0.7253	164.72	0.4356	0.5853	162.24
h=1	P- Likelihood	0.5538	0.5813	287.89	0.4059	0.3889	271.89
	G- Speckman	0.6736	0.5522	289.23	0.4036	0.3872	271.23
	Backftting	0.5681	0.6265	289.57	0.4014	0.3855	271.13
h=0.5	P- Likelihood	0.5044	0. 3332	402.63	0.2636	0. 1812	388.34
	G- Speckman	0.5338	0.3861	401.69	0.2676	0. 1852	388.58
	Backftting	0.5143	0.3920	403.71	0.2634	0. 1867	388.89
h=0.2	P- Likelihood	0.4734	0.1245	520.32	0.3136	0.0433	507.57
	G- Speckman	0.4912	0.1779	521.66	0.3255	0.0647	507.48
	Backftting	0.4840	0.1511	520.97	0.3269	0.0689	507.61

Table (6). Simulation Results for n = 1000

Criterion / Bandwidth and model		n = 1000					
		Classical Estimation			Bayesian Estimation		
		AMSE ($\hat{\beta}$)	AMSE ($\hat{m}$)	Mean Of Deviance	AMSE ($\hat{\beta}$)	AMSE ($\hat{m}$)	Mean Of Deviance
h=2	P- Likelihood	0.3354	0.4647	363.61	0.1943	0.3121	359.27
	G- Speckman	0.3360	0.4641	364.43	0.1977	0.3108	359.41
	Backftting	0.3434	0.4699	364.24	0.2039	0.3183	359.71
h=1	P- Likelihood	0.2011	0.3011	587.70	0. 0117	0.3011	584.76
	G- Speckman	0.2121	0.3206	589.40	0.0112	0.3216	584.43
	Backftting	0.2162	0.3223	589.34	0.0132	0.3223	584.96
h=0.5	P- Likelihood	0.1429	0. 1510	802.12	0.1029	0. 1113	800.12
	G- Speckman	0.1456	0.1612	801.60	0.1056	0.1119	800.60
	Backftting	0.1460	0.1633	803. 11	0.1060	0.1143	800. 91
h=0.2	P- Likelihood	0.0617	0.0715	1020.37	0.0358	0.0322	1018.37
	G- Speckman	0.0631	0.0734	1021.63	0.0326	0.0329	1018.63
	Backftting	0.0635	0.0747	1022.97	0.0414	0.0330	1020.97

Table (7). Simulation Results for n =2000

Criterion / Bandwidth and model		n = 2000					
		Classical Estimation			Bayesian Estimation		
		AMSE ($\hat{\beta}$)	AMSE ($\hat{m}$)	Mean Of Deviance	AMSE ($\hat{\beta}$)	AMSE ($\hat{m}$)	Mean Of Deviance
h=2	P- Likelihood	0.3354	0.4647	363.61	0.1943	0.3121	359.27
	G- Speckman	0.3360	0.4641	364.43	0.1977	0.3108	359.41
	Backftting	0.3434	0.4699	364.24	0.2039	0.3183	359.71
h=1	P- Likelihood	0.2011	0.3011	587.70	0. 0117	0.3011	584.76
	G- Speckman	0.2121	0.3206	589.40	0.0112	0.3216	584.43
	Backftting	0.2162	0.3223	589.34	0.0132	0.3223	584.96
h=0.5	P- Likelihood	0.1429	0. 1510	802.12	0.1029	0. 1113	800.12
	G- Speckman	0.1456	0.1612	801.60	0.1056	0.1119	800.60
	Backftting	0.1460	0.1633	803. 11	0.1060	0.1143	800. 91
h=0.2	P- Likelihood	0.0617	0.0715	1020.37	0.0358	0.0322	1018.37
	G- Speckman	0.0631	0.0734	1021.63	0.0326	0.0329	1018.63
	Backftting	0.0635	0.0747	1022.97	0.0414	0.0330	1020.97

From the results we can conclude that

1. The Bayesian estimation for the GPLM outperforms the classical estimation for all sample sizes under the square error loss function. It gives high efficient estimators with smallest AMSE $(\hat{\beta})$, AMSE $(\hat{m})$ and Mean of Deviance.
2. The Bayesian estimation for the GPLM using the back-fitting method outperforms the profile likelihood method and the generalized Speckman method for n=50 and n=100.
3. The Bayesian estimation for GPLM using the profile likelihood method and the generalized Speckman method outperforms the back-fitting method for n=1000 and n=2000.

5. DISCUSSIONS

In this article, we introduce a new Bayesian regression model called the Bayesian generalized partial linear model which extends the generalized partial linear model (GPLM). Bayesian estimation and inference of parameters for (GPLM) have been considered using some multivariate conjugate prior distributions under the square error loss function.

Simulation study is conducted to evaluate the performance of the proposed Bayesian estimators. Also, the simulation study is used to compare the proposed technique with classical approaches. The simulation study is based on 500 Monte Carlo replications. Different sample sizes have been used ranging from small, moderate, to large sizes. In specific sample sizes are fixed at n=50, n=100, n=200, n=500, n=100, and n=2000. Also, different bandwidth parameters are used, namely, h=2, h=1, h=0.5, and h=0.2.

From the results of the simulation study, we can conclude the following. First, the Bayesian estimation for the GPLM outperforms the classical estimation for all sample sizes under the square error loss function. It gives high efficient estimators with smallest AMSE $(\hat{\beta})$, AMSE $(\hat{m})$ and Mean of Deviance. Second, the Bayesian estimation for the GPLM using the back-fitting method outperforms the profile likelihood method and the generalized Speckman method for n=50 and n=100. The Bayesian estimation for GPLM using the profile likelihood method and the generalized Speckman method outperforms the back-fitting method for n=1000 and n=2000. Finally, The Bayesian estimation of parameters for the GPLM gives small values of AMSE $(\hat{\beta})$, AMSE $(\hat{m})$ comparable to the classical estimation for all sample sizes.

REFERENCES

1. Bornkamp, B., and Ickstadt, K. (2009) Bayesian nonparametric estimation of continuous monotone functions with applications to dose–response analysis, Biometrics, 65, 198–205.
2. Bouaziz, O., Geffray, S., and Lopez, O. (2015) Semiparametric inference for the recurrent events process by means of a single-index model, Journal of Theoretical and Applied Statistics, 49 (2), 361–385.
3. Burda, M., Hardel, W., Müller, M., and Weratz, A. (1994) Semiparamrtric analysis of German east-west migration: facts and theory, Journal of Applied Econometrics, 13 (5), 525-542.
4. Guo, X., Niu, G., Yang, Y., and Xu,W. (2015) Empirical likelihood for single index model with missing covariates at random, 49 (3), 588–601.
5. Holmes, C.C., and Heard, N. A. (2003) Generalized monotonic regression using random change points, Statistics in Medicine, 22, 623–638.
6. Johnson, M.S. (2007) Modeling dichotomous item responses with free-knot splines, Journal of Computational Statistics& Data Analysis, 51, 4178–4192.
7. Lang, S., and Brezger,A. (2004), Bayesian P-splines, Journal of Computational and Graphical Statistics, 13, 183–212.
8. Meyer, M.C., Hackstadt, A. J., and Hoeting, J. A. (2011) Bayesian estimation and inference for generalized partial linear models using shape-restricted splines, Journal of Nonparametric Statistics, 23, 867-884.
9. Müller, M. (2001) Estimation and testing in generalized partial linear models: a comparative study, Statistics and Computing, 11, 299-309.
10. Murphy, S.A., and Van Der Vaart, A.W. (2000) On profile likelihood, Journal of the American Statistical Association, 95, 449-485.
11. Neelon, B., and Dunson, D.B. (2004) Bayesian isotonic regression and trend analysis, Biometrics, 60, 498–406.
12. Peng, H. (2003) Efficient estimation in semi- parametric generalized linear models, Proceedings of JSM 2003.
13. Peng, H., and Wang, X., (2004) Moment estimation in semi-parametric generalized linear models, Nonparametric Statistics, 1-22.

14. Perron, F., and Mengersen, K. (2001) Bayesian nonparametric modeling using mixtures of triangular distributions, Biometrics, 57, 518–528.

15. Powel, L. (1994) Estimation of semi-parametric regression models, Handbook of Econometrics, 4, Ch. 41.

16. Brezger, A., and Steiner,W.J. (2008) Monotonic regression based on Bayesian P-splines: an application to estimating price response functions from store-level scanner data , Journal of Business & Economic Statistics, 26, 91–104.

17. Ramgopal, P., Laud, P.W., and Smith, A.F.M. (1993) Nonparametric Bayesian bioassay with prior constraints on the shape of the potency curve, Biometrika, 80, 489–498.

18. Ruppert, D., Wang, M.P., and Caroll, R.J., (2003) Semi-parametric regression, Cambridge University Press, London.

19. Severini, T.A., and Staniswalis, J.G., (1994) Quasi-likelihood estimation in semi-parametric models, Journal of the American Statistical Association, 89, 501-511

20. Severini, T.A, and Wong, W.H., (1992) Profile likelihood and conditionally parametric models, The Annals of Statistics, 20, 1768-1802.

21. Shively, T.S., Sager, T.W., and Walker, S.G. (2009) A Bayesian approach to non-parametric monotone function estimation, Journal of the Royal Statistical Society B, 71, 159–175.

22. Sperlich, S., Hardle, W., and Aydinli, G., (2006) The art of semi-parametric, Physical-Verlag, Germany.

23. Wang, X. (2008) Bayesian free-knot monotone cubic spline regression, Journal of Computational and Graphical Statistics, 17, 373–387.

24. XploRe (2000), "XploRe-the interactive statistical computing environment" WWW: http://www.xplore-stat.de.

25. Zhang, J., Wang, X., Yu, Y., and Gai, Y., (2010) Estimation and variable selection in partial linear single index models with error-prone linear covariates, Journal of Theoretical and Applied Statistics, 48 (5), 1048–1070

Chapter 3

Akash Distribution and Its Applications

Rama Shanker

Department of Statistics, Eritrea Institute of Technology, Asmara, Eritrea

ABSTRACT

A new one parameter lifetime distribution named "Akash distribution" for modeling lifetime data has been introduced. Some important mathematical properties of the proposed distribution including its shape, moments, skewness, kurtosis, hazard rate function, mean residual life function, stochastic ordering, mean deviations, order statistics, Bonferroni and Lorenz curves, Renyi entropy measure, stress-strength reliability have been discussed. The condition under which Akash distribution is over-dispersed, equi-dispersed, and under-dispersed are presented along with the conditions under which exponential and Lindley distributions are over-dispersed, equi-dispersed and under-dispersed. The estimation of its parameter has been discussed using maximum likelihood estimation and method of moments. The usefulness and the applicability of the proposed distribution have been discussed and illustrated with two real lifetime data sets from medical science and engineering.

Keywords: Lifetime distribution, Moments, Hazard rate function, Mean residual life function, Mean deviations, Order statistics, Estimation of parameter, Goodness of fit

1. INTRODUCTION

The modeling and analyzing lifetime data are crucial in many applied sciences including medicine, engineering, insurance and finance, amongst others. There are a number of continuous distributions for modeling lifetime data such as exponential, Lindley, gamma, lognormal, and Weibull and their generalizations. The exponential, Lindley and the Weibull distributions are more popular than the gamma and the lognormal distributions because the survival functions of the gamma and the lognormal distributions cannot be expressed in closed forms and both require numerical integration. Though each of exponential and Lindley distributions has one parameter, the Lindley distribution has one advantage over the exponential distribution that the exponential distribution has constant hazard rate whereas the Lindley distribution has monotonically decreasing hazard rate.

The probability density function (p.d.f.) and the cumulative distribution function (c.d.f.) of Lindley (1958) distribution are given by

$$f(x;\theta)=\frac{\theta^2}{\theta+1}(1+x)e^{-\theta x} \quad ;x>0,\theta>0 \tag{1.1}$$

$$F(x;\theta)=1-\left[1+\frac{\theta x}{\theta+1}\right]e^{-\theta x} \quad ;x>0,\theta>0 \tag{1.2}$$

The density (1.1) is a two-component mixture of an exponential distribution having scale parameter θ and a gamma distribution having shape parameter 2 and scale parameter θ with their mixing proportions $\frac{\theta}{\theta+1}$ and $\frac{1}{\theta+1}$ respectively. A detailed study about its various mathematical properties, estimation of parameter and application showing the superiority of Lindley distribution over exponential distribution for the waiting times before service of the bank customers has been done by Ghitany *et al* (2008). The Lindley distribution has been generalized, extended, modified and its detailed applications in reliability and other fields of knowledge by different researchers including Hussain (2006), Zakerzadeh and Dolati (2009), Nadarajah *et al* (2011), Deniz and Ojeda (2011), Bakouch et al (2012), Shanker and Mishra (2013 a, 2013 b), Shanker *et al* (2013), Elbatal *et al*(2013), Ghitany *et al* (2013), Merovci (2013), Liyanage and Pararai (2014), Ashour and Eltehiwy (2014), Oluyede and Yang (2014), Singh *et al* (2014), Sharma *et al* (2015), Shanker *et al* (2015), Alkarni (2015), Pararai *et al*(2015), Abouammoh *et al* (2015) are some among others.

Although the Lindley distribution has been used to model lifetime data by many researchers and Hussain (2006) has shown that the Lindley distribution is important for studying stress-strength reliability modeling, there are many situations in the modeling of real lifetime data where the Lindley distribution may not be suitable from a theoretical or applied point of view. Therefore, to obtain a new distribution which is flexible than the Lindley distribution for modeling lifetime data in reliability and in terms of its hazard rate shapes, we introduced a new distribution by considering a two- considering a two-component mixture of an exponential distribution having scale parameter θ and a gamma distribution having shape parameter 3 and scale parameter θ with their mixing proportions $\frac{\theta^2}{\theta^2+2}$ and $\frac{2}{\theta^2+2}$ respectively.

The probability density function (p.d.f.) of a new one parameter lifetime distribution can be introduced as

$$f(x;\theta)=\frac{\theta^3}{\theta^2+2}\left(1+x^2\right)e^{-\theta x} \quad ;x>0,\ \theta>0 \tag{1.3}$$

We would call this distribution, "Akash distribution. This distribution can be easily expressed as a mixture of exponential (θ) and gamma $(3,\theta)$ with their mixing proportions $\frac{\theta^2}{\theta^2+2}$ and $\frac{2}{\theta^2+2}$ respectively. We have

$$f(x,\theta)=p\,g_1(x)+(1-p)\,g_2(x)$$

where

$$p=\frac{\theta^2}{\theta^2+2},\ g_1(x)=\theta e^{-\theta x},\ \text{and}\ g_2(x)=\frac{\theta^3x^2e^{-\theta x}}{2}.$$

The corresponding cumulative distribution function (c.d.f.) of (1.3) is given by

$$F(x)=1-\left[1+\frac{\theta x(\theta x+2)}{\theta^2+2}\right]e^{-\theta x};\ x>0,\theta>0 \tag{1.4}$$

The graphs of the p.d.f. and the c.d.f. of Lindley and Akash distributions for different values of θ are shown in **figures 1 and 2.**

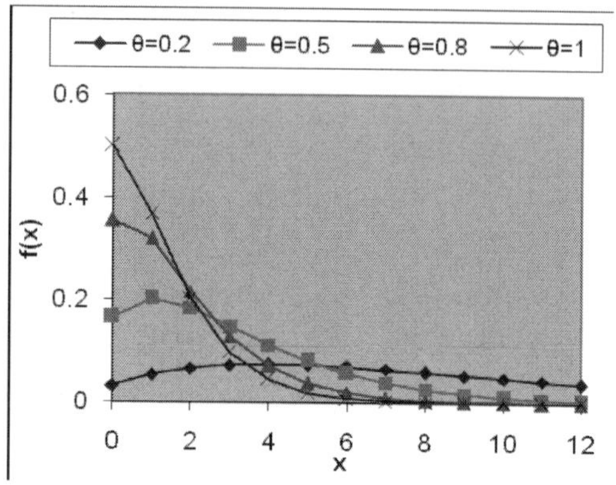
θ=0.2
θ=0.5
θ=0.8
θ=1
0.6
0.4
0.2
0
f(x)
0
2
4
6
8
10
12
x

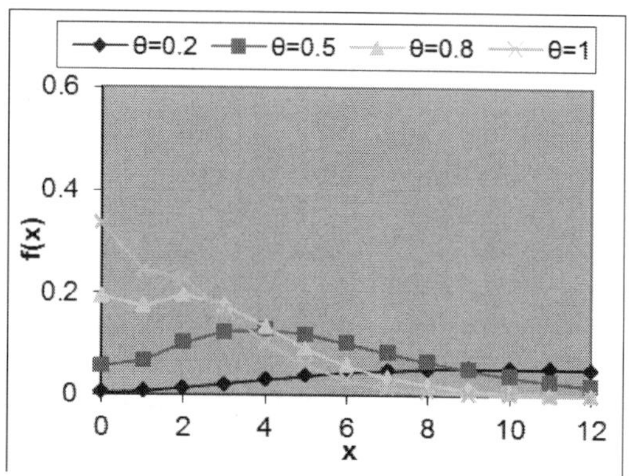
θ=0.2
θ=0.5
θ=0.8
θ=1
0.6
0.4
0.2
0
f(x)
0
2
4
6
8
10
12
x

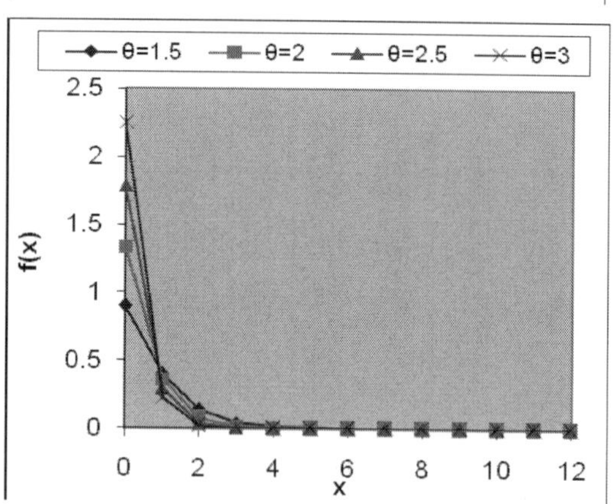
θ=1.5
θ=2
θ=2.5
θ=3
2.5
2
1.5
1
0.5
0
f(x)
0
2
4
6
8
10
12
x

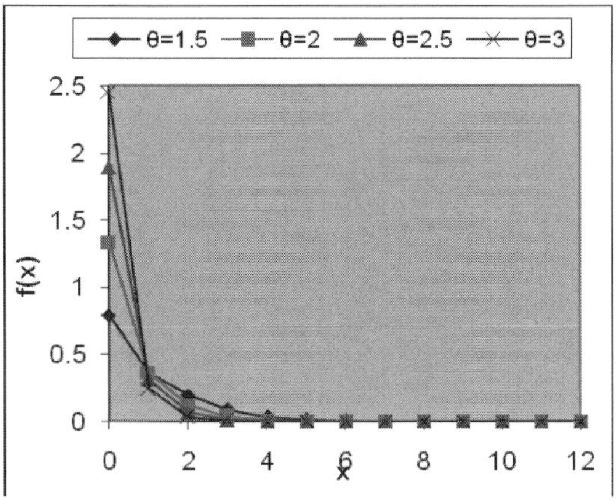

Figure 1. Graphs of the pdf of Lindley and Akash distributions for different values of parameter . Left hand side graphs are for Lindley distribution and right hand sides graph are for Akash distribution θ

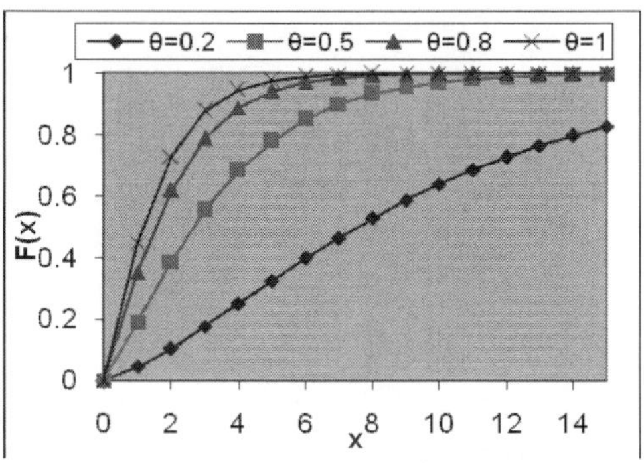

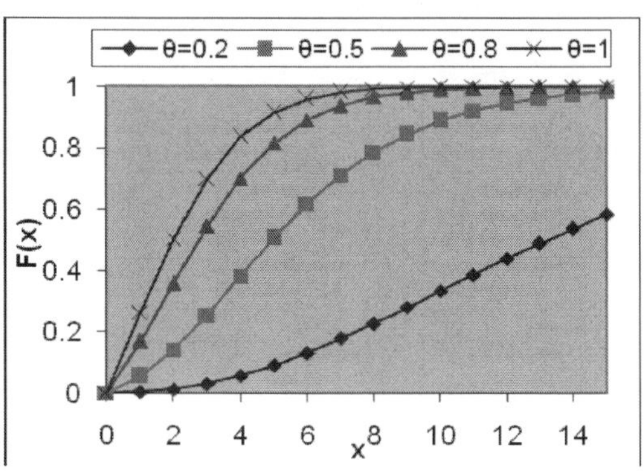

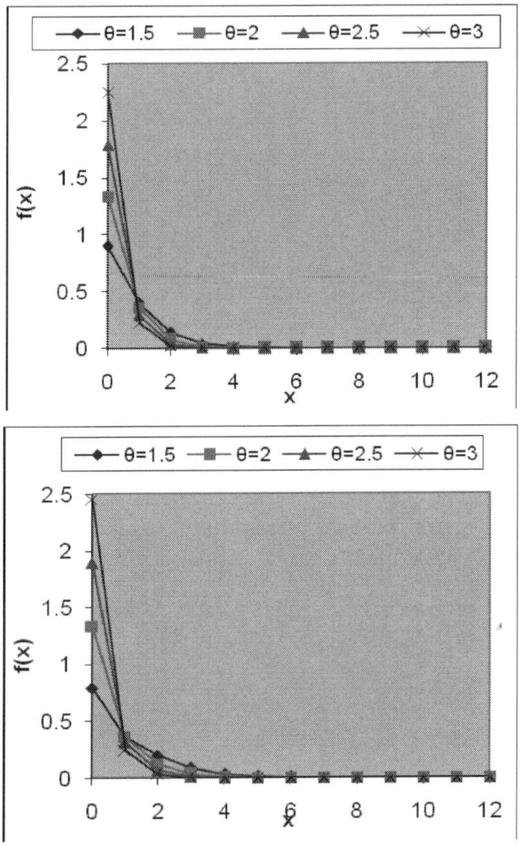

Figure 2. Graphs of the cdf of Lindley and Akash distribution for different values of parameter . Left hand side graphs are for Lindley distribution and right hand side graphs are for Akash distribution θ

2. MOMENTS AND RELATED MEASURES

The r the moment about origin of Akash distribution (1.3) has been obtained as

$$\mu_r' = \frac{r!\left[\theta^2+(r+1)(r+2)\right]}{\theta^r\left(\theta^2+2\right)} \quad ;r=1,2,3,...$$

and so the first four moments about origin as

$$\mu_1' = \frac{\theta^2 + 6}{\theta(\theta^2 + 2)}, \quad \mu_2' = \frac{2(\theta^2 + 12)}{\theta^2(\theta^2 + 2)};$$

$$\mu_3' = \frac{6(\theta^2 + 20)}{\theta^3(\theta^2 + 2)}, \quad \mu_4' = \frac{24(\theta^2 + 30)}{\theta^4(\theta^2 + 2)}$$

Thus the moments about mean of Akash distribution are obtained as

$$\mu_2 = \frac{\theta^4 + 16\theta^2 + 12}{\theta^2(\theta^2 + 2)^2}$$

$$\mu_3 = \frac{2(\theta^6 + 30\theta^4 + 36\theta^2 + 24)}{\theta^3(\theta^2 + 2)^3}$$

$$\mu_4 = \frac{3(3\theta^8 + 128\theta^6 + 408\theta^4 + 576\theta^2 + 240)}{\theta^4(\theta^2 + 2)^4}$$

The coefficient of variation $(C.V)$, coefficient of skewness $(\sqrt{\beta_1})$, coefficient of kurtosis (β_2) and index of dispersion (γ) of Akash distribution are thus obtained as

$$C.V = \frac{\sigma}{\mu_1'} = \frac{\sqrt{\theta^4 + 16\theta^2 + 12}}{\theta^2 + 6}$$

$$\sqrt{\beta_1} = \frac{\mu_3}{\mu_2^{3/2}} = \frac{2(\theta^6 + 30\theta^4 + 36\theta^2 + 24)}{(\theta^4 + 16\theta^2 + 12)^{3/2}}$$

$$\beta_2 = \frac{\mu_4}{\mu_2^2} = \frac{3(3\theta^8 + 128\theta^6 + 408\theta^4 + 576\theta^2 + 240)}{(\theta^4 + 16\theta^2 + 12)^2}$$

$$\gamma = \frac{\sigma^2}{\mu_1'} = \frac{\theta^4 + 16\theta^2 + 12}{\theta(\theta^2 + 2)(\theta^2 + 6)}$$

It can be easily shown that Akash distribution is over-dispersed $(\mu < \sigma^2)$, equi-dispersed $(\mu = \sigma^2)$ and under-dispers $(\mu > \sigma^2)$ for $\theta < (=) > \theta^* = 1.515400063$. It would be recalled that Lindley distribution is over-dispersed $(\mu < \sigma^2)$, equi-dispersed $(\mu = \sigma^2)$ and under-dispers $(\mu > \sigma^2)$ for $\theta < (=) > \theta^* = 1.170086487$

while exponential distribution is over-dispersed $\left(\mu<\sigma^2\right)$, equi-dispersed $\left(\mu=\sigma^2\right)$ and under-dispersed $\left(\mu>\sigma^2\right)$ for $\theta<(=)>\theta^*=1$.

3. MOMENT GENERATING FUNCTION

The moment generating function of Akash distribution (1.3) are obtained as

$$\begin{aligned}M_X(t)&=\frac{\theta^3}{\theta^2+2}\int_0^\infty e^{-(\theta-t)}\left(1+x^2\right)dx\\&=\frac{\theta^3}{\theta^2+2}\left[\frac{1}{\theta-t}+\frac{2}{(\theta-t)^3}\right]\\&=\frac{\theta^3}{\theta^2+2}\left[\frac{1}{\theta}\sum_{k=0}^{\infty}\left(\frac{t}{\theta}\right)^k+\frac{2}{\theta^3}\sum_{k=0}^{\infty}\binom{k+2}{k}\left(\frac{t}{\theta}\right)^k\right]\\&=\sum_{k=0}^{\infty}\frac{\theta^2+(k+1)(k+2)}{\theta^2+2}\left(\frac{t}{\theta}\right)^k\end{aligned}$$

4. HAZARD RATE FUNCTION AND MEAN RESIDUAL LIFE FUNCTION

Let X be a continuous random variable with p.d.f. $f(x)$ and c.d.f. $F(x)$. The hazard rate function (also known as the failure rate function) and the mean residual life function of X are respectively defined as

$$h(x)=\lim_{\Delta x\to 0}\frac{P(X<x+\Delta x|X>x)}{\Delta x}=\frac{f(x)}{1-F(x)} \tag{4.1}$$

And

$$m(x)=E\left[X-x|X>x\right]=\frac{1}{1-F(x)}\int_x^\infty\left[1-F(t)\right]\,dt \tag{4.2}$$

The hazard rate function, $h(x)$ and the mean residual life function, $m(x)$ of Lindley distribution are given by

$$h(x)=\frac{\theta^2(1+x)}{(\theta+1)+\theta x} \tag{4.3}$$

And

$$m(x)=\frac{\theta+2+\theta x}{\theta(\theta+1+\theta x)} \tag{4.4}$$

The corresponding hazard rate function, $h(x)$ and the mean residual life function, $m(x)$ of the Akash distribution are given by

$$h(x)=\frac{\theta^3\left(1+x^2\right)}{\theta x(\theta x+2)+\left(\theta^2+2\right)} \tag{4.5}$$

And

$$m(x)=\frac{\theta^2+2}{\left[\theta x(\theta x+2)+\left(\theta^2+2\right)\right]e^{-\theta x}}\int_x^\infty\left[\frac{\theta t(\theta t+2)+\left(\theta^2+2\right)}{\theta^2+2}\right]e^{-\theta t}dt=\frac{\theta^2x^2+4\theta x+\left(\theta^2+6\right)}{\theta\left[\theta x(\theta x+2)+\left(\theta^2+2\right)\right]} \tag{4.6}$$

It can be easily verified that $h(0)=\frac{\theta^3}{\theta^2+2}=f(0)$ and $m(0)=\frac{\theta^2+6}{\theta\left(\theta^2+2\right)}=\mu_1'$. It is also obvious from the graphs of $h(x)$ and $m(x)$ that $h(x)$ is an increasing function of x, and θ, whereas $m(x)$ is a decreasing function of x, and θ. The hazard rate function and the mean residual life function of the Akash distribution show its flexibility over Lindley distribution and exponential distribution.

The graphs of the hazard rate function and mean residual life function of Lindley and Akash distributions are shown in figures 3 and 4.

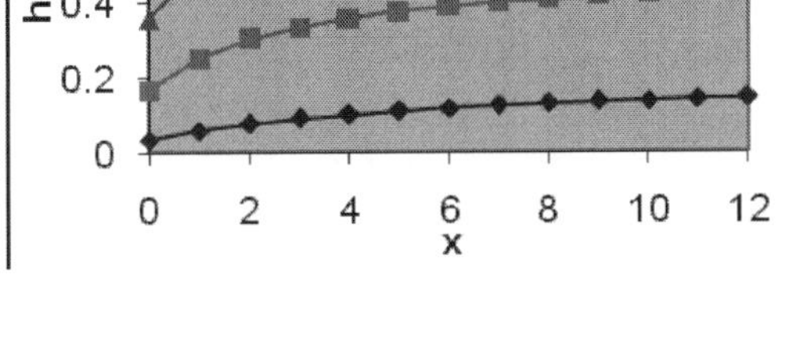

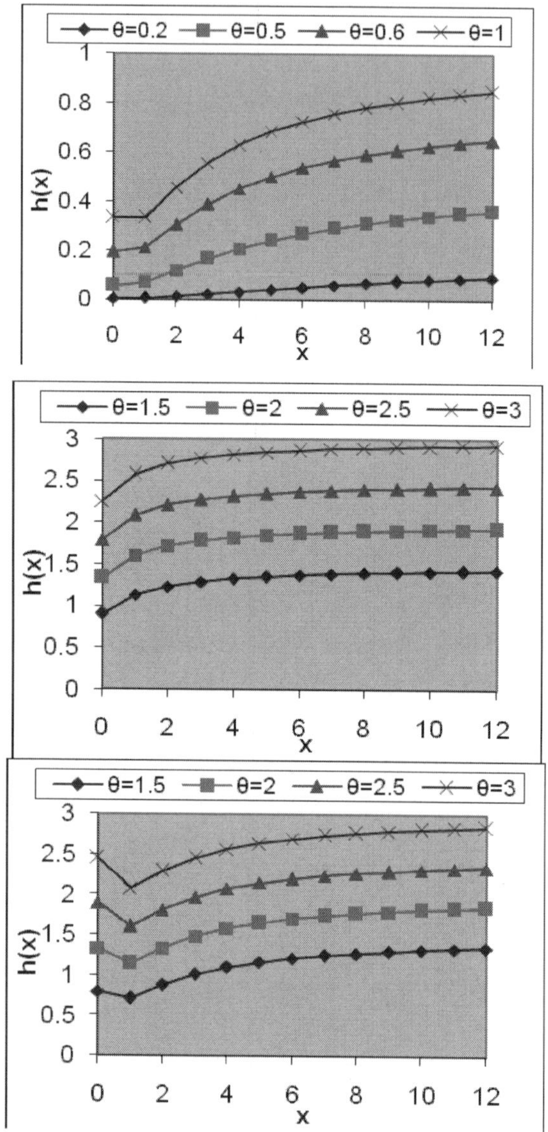

Figure 3. Graphs of hazard rate function of Lindley and Akash distributions for different values of parameter . Left hand side graphs are for Lindley distribution and right hand side graphs are for Akash distribution θ

5. STOCHASTIC ORDERINGS

Stochastic ordering of positive continuous random variables is an important tool for judging their comparative behavior. A random variable X is said to be smaller than a random variable Y in the

(i) stochastic order $(X \le_{st} Y)$ if $F_X(x) \ge F_Y(x)$ for all x

(ii) hazard rate order $(X \le_{hr} Y)$ if $h_X(x) \ge h_Y(x)$ for all x

(iii) mean residual life order $(X \le_{mrl} Y)$ if $m_X(x) \le m_Y(x)$ for all x

(iv) likelihood ratio order $(X \le_{lr} Y)$ if $\dfrac{f_X(x)}{f_Y(x)}$ decreases in x.

The following results due to Shaked and Shanthikumar (1994) are well known for establishing stochastic ordering of distributions

$$\begin{array}{c} X \le_{lr} Y \Rightarrow X \le_{hr} Y \Rightarrow X \le_{mrl} Y \\ \Downarrow \\ X \le_{st} Y \end{array} \tag{5.1}$$

The Akash distribution is ordered with respect to the strongest 'likelihood ratio' ordering as shown in the following theorem:

Theorem: Let X ~ Akash distributon (θ_1) and Y ~ Akash distribution (θ_2). If $\theta_1 \ge \theta_2$, then $X \le_{lr} Y$ and hence $X \le_{hr} Y$, $X \le_{mrl} Y$ and $X \le_{st} Y$.

Proof: We have

$$\frac{f_X(x)}{f_Y(x)} = \frac{\theta_1^3\left(\theta_2^2+2\right)}{\theta_2^3\left(\theta_1^2+2\right)} e^{-(\theta_1-\theta_2)x} \quad ; x>0$$

Now

$$\log\frac{f_X(x)}{f_Y(x)} = \log\left[\frac{\theta_1^3\left(\theta_2^2+2\right)}{\theta_2^3\left(\theta_1^2+2\right)}\right] - \left(\theta_1-\theta_2\right)x$$

This gives $\frac{d}{dx}\log\frac{f_X(x)}{f_Y(x)} = -(\theta_1 - \theta_2)$

Thus for $\theta_1 \geq \theta_2$, $\frac{d}{dx}\log\frac{f_X(x)}{f_Y(x)} < 0$. This means that $X \leq_{lr} Y$ and hence $X \leq_{hr} Y$, $X \leq_{mrl} Y$ and $X \leq_{st} Y$.

6. MEAN DEVIATIONS

The amount of scatter in a population is measured to some extent by the totality of deviations usually from mean and median. These are known as the mean deviation about the mean and the mean deviation about the median defined by

$\delta_1(X) = \int_0^{\infty} |x-\mu| f(x)dx$ and $\delta_2(X) = \int_0^{\infty} |x-M| f(x)dx$, respectively,

where $\mu = E(X)$ and $M = \text{Median }(X)$. The measures $\delta_1(X)$ and $\delta_2(X)$ can be calculated using the relationships

$$\begin{aligned}\delta_1(X) &= \int_0^{\mu}(\mu - x)f(x)dx + \int_{\mu}^{\infty}(x-\mu)f(x)dx \\ &= \mu F(\mu) - \int_0^{\mu} xf(x)dx - \mu[1-F(\mu)] + \int_{\mu}^{\infty} xf(x)dx \\ &= 2\mu F(\mu) - 2\mu + 2\int_{\mu}^{\infty} xf(x)dx \\ &= 2\mu F(\mu) - 2\int_0^{\mu} xf(x)dx \end{aligned} \tag{6.1}$$

And

$$\begin{aligned}\delta_2(X) &= \int_0^{M}(M - x)f(x)dx + \int_{M}^{\infty}(x-M)f(x)dx \\ &= MF(M) - \int_0^{M} xf(x)dx - M[1-F(M)] + \int_{M}^{\infty} xf(x)dx \\ &= -\mu + 2\int_{M}^{\infty} xf(x)dx \\ &= \mu - 2\int_0^{M} xf(x)dx \end{aligned} \tag{6.2}$$

Using p.d.f. (1.3) and expression for the mean of Akash distribution, we get

$$\int_0^{\mu} x f(x)dx$$

$$= \mu - \frac{\left\{\theta^3\left(\mu^3+\mu\right)+\theta^2\left(3\mu^2+1\right)+6\left(\theta\mu+1\right)\right\}e^{-\theta\mu}}{\theta\left(\theta^2+2\right)} \tag{6.3}$$

$$\int_0^{M} x f(x)dx$$

$$= \mu - \frac{\left\{\theta^3\left(M^3+M\right)+\theta^2\left(3M^2+1\right)+6\left(\theta M+1\right)\right\}e^{-\theta M}}{\theta\left(\theta^2+2\right)} \tag{6.4}$$

Using expressions from (6.1), (6.2), (6.3), and (6.4), the mean deviation about mean, $\delta_1(X)$ and the mean deviation about median, $\delta_2(X)$ of Akash distribution are obtained as

$$\delta_1(X) = \frac{2\left\{\theta^2\left(\mu^2+1\right)+2\left(2\theta\mu+3\right)\right\}e^{-\theta\mu}}{\theta\left(\theta^2+2\right)} \tag{6.5}$$

$$\delta_2(X) = \frac{2\left\{\theta^3\left(M^3+M\right)+\theta^2\left(3M^2+1\right)+6\left(\theta M+1\right)\right\}e^{-\theta M}}{\theta\left(\theta^2+2\right)} - \mu \tag{6.6}$$

7. ORDER STATISTICS

Let $X_1, X_2, \ldots, X_n$ be a random sample of size n from Akash distribution (1.3). Let $X_{(1)} < X_{(2)} < \ldots < X_{(n)}$ denote the corresponding order statistics. The p.d.f. and the c.d.f. of the k th order statistic, say $Y = X_{(k)}$ are given by

$$f_Y(y) = \frac{n!}{(k-1)!(n-k)!} F^{k-1}(y)\left\{1-F(y)\right\}^{n-k} f(y)$$

$$= \frac{n!}{(k-1)!(n-k)!} \sum_{l=0}^{n-k} \binom{n-k}{l} (-1)^l F^{k+l-1}(y) f(y)$$

and

$$F_Y(y)=\sum_{j=k}^{n}\binom{n}{j}F^j(y)\{1-F(y)\}^{n-j}$$

$$=\sum_{j=k}^{n}\sum_{l=0}^{n-j}\binom{n}{j}\binom{n-j}{l}(-1)^l F^{j+l}(y),$$

respectively, for $k=1,2,3,\ldots,n$.

Thus, the p.d.f. and the c.d.f of k th order statistics are given by

$$f_Y(y)=\frac{n!\theta^3(1+x^2)e^{-\theta x}}{(\theta^2+2)(k-1)!(n-k)!}\sum_{l=0}^{n-k}\binom{n-k}{l}(-1)^l\left[1-\frac{\theta x(\theta x+2)+(\theta^2+2)}{\theta^2+2}e^{-\theta x}\right]^{k+l-1}$$

and

$$F_Y(y)=\sum_{j=k}^{n}\sum_{l=0}^{n-j}\binom{n}{j}\binom{n-j}{l}(-1)^l\left[1-\frac{\theta x(\theta x+2)+(\theta^2+2)}{\theta^2+2}e^{-\theta x}\right]^{j+l}$$

8. BONFERRONI AND LORENZ CURVES

The Bonferroni and Lorenz curves (Bonferroni, 1930) and Bonferroni and Gini indices have applications not only in economics to study income and poverty, but also in other fields like reliability, demography, insurance and medicine. The Bonferroni and Lorenz curves are defined as

$$B(p)=\frac{1}{p\mu}\int_0^q x f(x)dx=\frac{1}{p\mu}\left[\int_0^\infty x f(x)dx-\int_q^\infty x f(x)dx\right]=\frac{1}{p\mu}\left[\mu-\int_q^\infty x f(x)dx\right] \tag{8.1}$$

And

$$L(p)=\frac{1}{\mu}\int_0^q x f(x)dx=\frac{1}{\mu}\left[\int_0^\infty x f(x)dx-\int_q^\infty x f(x)dx\right]=\frac{1}{\mu}\left[\mu-\int_q^\infty x f(x)dx\right] \tag{8.2}$$

respectively or equivalently

$$B(p)=\frac{1}{p\mu}\int_0^p F^{-1}(x)dx \tag{8.3}$$

$$\text{and } L(p)=\frac{1}{\mu}\int_0^p F^{-1}(x)dx \tag{8.4}$$

respectively, where $\mu = E(X)$ and $q = F^{-1}(p)$.

The Bonferroni and Gini indices are thus defined as

$$B = 1 - \int_0^1 B(p)\,dp \tag{8.5}$$

$$\text{and } G = 1 - 2\int_0^1 L(p)\,dp \tag{8.6}$$

respectively.

Using p.d.f. (1.3), we get

$$\int_q^\infty x f(x)\,dx = \frac{\left\{\theta^3\left(q^3+q\right)+\theta^2\left(3q^2+1\right)+6\left(\theta q+1\right)\right\}e^{-\theta q}}{\theta\left(\theta^2+2\right)} \tag{8.7}$$

Now using equation (8.7) in (8.1) and (8.2), we get

$$B(p) = \frac{1}{p}\left[1-\frac{\left\{\theta^3\left(q^3+q\right)+\theta^2\left(3q^2+1\right)+6\left(\theta q+1\right)\right\}e^{-\theta q}}{\theta^2+6}\right] \tag{8.8}$$

$$\text{and } L(p) = 1-\frac{\left\{\theta^3\left(q^3+q\right)+\theta^2\left(3q^2+1\right)+6\left(\theta q+1\right)\right\}e^{-\theta q}}{\theta^2+6} \tag{8.9}$$

Now using equations (8.8) and (8.9) in (8.5) and (8.6), the Bonferroni and Gini indices are obtained as

$$B = 1-\frac{\left\{\theta^3\left(q^3+q\right)+\theta^2\left(3q^2+1\right)+6\left(\theta q+1\right)\right\}e^{-\theta q}}{\theta^2+6} \tag{8.10}$$

$$G = -1+\frac{2\left\{\theta^3\left(q^3+q\right)+\theta^2\left(3q^2+1\right)+6\left(\theta q+1\right)\right\}e^{-\theta q}}{\theta^2+6} \tag{8.11}$$

9. RENYI ENTROPY

An entropy of a random variable X is a measure of variation of uncertainty. A popular entropy measure is Renyi entropy (1961). If X is a continuous random variable having probability density function $f(\cdot)$, then Renyi entropy is defined as

$$T_R(\gamma)=\frac{1}{1-\gamma}\log\left\{\int f^{\gamma}(x)dx\right\}$$

where $\gamma>0$ and $\gamma\neq 1$.

Thus, the Renyi entropy for the Akash distribution (1.3) is obtained as

$$T_R(\gamma)=\frac{1}{1-\gamma}\log\left[\int_0^{\infty}\frac{\theta^{3\gamma}}{\left(\theta^2+2\right)^{\gamma}}\left(1+x^2\right)^{\gamma}e^{-\theta\gamma x}dx\right]$$

$$=\frac{1}{1-\gamma}\log\left[\int_0^{\infty}\frac{\theta^{3\gamma}}{\left(\theta^2+2\right)^{\gamma}}\sum_{j=0}^{\infty}\binom{\gamma}{j}\left(x^2\right)^{j}e^{-\theta\gamma x}dx\right]$$

$$=\frac{1}{1-\gamma}\log\left[\int_0^{\infty}\frac{\theta^{3\gamma}}{\left(\theta^2+2\right)^{\gamma}}\sum_{j=0}^{\infty}\binom{\gamma}{j}x^{2j}e^{-\theta\gamma x}dx\right]$$

$$=\frac{1}{1-\gamma}\log\left[\sum_{j=0}^{\infty}\binom{\gamma}{j}\frac{\theta^{3\gamma}}{\left(\theta^2+2\right)^{\gamma}}\int_0^{\infty}e^{-\theta\gamma x}x^{2j+1-1}dx\right]$$

$$=\frac{1}{1-\gamma}\log\left[\sum_{j=0}^{\infty}\binom{\gamma}{j}\frac{\theta^{3\gamma}}{\left(\theta^2+2\right)^{\gamma}}\frac{\Gamma(2j+1)}{\left(\theta\gamma\right)^{2j+1}}\right]$$

$$=\frac{1}{1-\gamma}\log\left[\sum_{j=0}^{\infty}\binom{\gamma}{j}\frac{\theta^{3\gamma-2j-1}}{\left(\theta^2+2\right)^{\gamma}}\frac{\Gamma(2j+1)}{\left(\gamma\right)^{2j+1}}\right]$$

10. STRESS-STRENGTH RELIABILITY

The stress- strength reliability describes the life of a component which has random strength X that is subjected to a random stress Y. When the stress applied to it exceeds the strength, the component fails instantly and the component will function satisfactorily till $X>Y$. Therefore, $R=P(Y<X)$ is a measure of component reliability and in statistical literature it is known as stress-strength parameter. It has wide applications in almost all areas of

knowledge especially in engineering such as structures, deterioration of rocket motors, static fatigue of ceramic components, aging of concrete pressure vessels etc.

Let X and Y be independent strength and stress random variables having Akash distribution (1.3) with parameter θ_1 and θ_2 respectively. Then the stress-strength reliability R is obtained as

$$R = P(Y < X) = \int_0^\infty P(Y < X \mid X = x) f_X(x)\,dx$$

$$= \int_0^\infty f(x;\theta_1)\, F(x;\theta_2)\,dx$$

$$= 1 - \frac{\theta_1^3 \begin{bmatrix} \theta_2^6 + 4\theta_1\theta_2^5 + 2(3\theta_1^2+4)\theta_2^4 \\ +2(2\theta_1^2+11)\theta_1\theta_2^3 + (\theta_1^4+22\theta_1^2+40)\theta_2^2 \\ +10(\theta_1^2+2)\theta_1\theta_2 + 2(\theta_1^2+2)\theta_1^2 \end{bmatrix}}{(\theta_1^2+2)(\theta_2^2+2)(\theta_1+\theta_2)^5}.$$

11. ESTIMATION OF PARAMETER

11.1. Maximum Likelihood Estimation

Let $(x_1, x_2, x_3, \dots, x_n)$ be a random sample from Akash distribution (1.3). The likelihood function, L of (1.3) is given by

$$L = \left(\frac{\theta^3}{\theta^2+2}\right)^n \prod_{i=1}^n \left(1+x_i^2\right) e^{-n\theta\bar{x}}$$

and so its log likelihood function is thus obtained as

$$\log L = n\log\left(\frac{\theta^3}{\theta^2+2}\right) + \sum_{i=1}^n \log\left(1+x_i^2\right) - n\theta\bar{x}$$

Now $\frac{d\log L}{d\theta} = \frac{3n}{\theta} - \frac{2n\theta}{\theta^2+2} - n\bar{x}$

where $\bar{x}$ is the sample mean.

The maximum likelihood estimate, $\hat{\theta}$ of θ is the solution of the equation $\frac{d \log L}{d\theta} = 0$ and so it can be obtained by solving the following non-linear equation

$$\bar{x}\theta^3 - \theta^2 + 2\bar{x}\theta - 6 = 0 \tag{11.1.1}$$

11.2. Method of Moment (MoM) Estimation

Equating the population mean of the Akash distribution to the corresponding sample mean, the method of moment (MOM) estimate, $\tilde{\theta}$, of θ is the same as given by equation (11.1.1).

12. APPLICATIONS OF AKASH DISTRIBUTION

The Akash distribution has been fitted to a number of data sets from medical science and engineering. In this section, we present the fitting of Akash distribution to two real data sets and compare its goodness of fit with the one parameter exponential and Lindley distributions.

Data set 1: The first data set represents the lifetime's data relating to relief times (in minutes) of 20 patients receiving an analgesic and reported by Gross and Clark (1975, P. 105). The data are as follows:

1.1, 1.4, 1.3, 1.7, 1.9, 1.8, 1.6, 2.2, 1.7, 2.7,
4.1, 1.8, 1.5, 1.2, 1.4, 3.0, 1.7, 2.3, 1.6, 2.0

Data set 2: The second data set is the strength data of glass of the aircraft window reported by Fuller *et al*(1994)

18.83, 20.80, 21.657, 23.03, 23.23, 24.05,
24.321, 25.50, 25.52, 25.80, 26.69, 26.77,
26.78, 27.05, 27.67, 29.90, 31.11, 33.20,
33.73, 33.76, 33.89, 34.76, 35.75, 35.91,
36.98, 37.08, 37.09, 39.58, 44.045, 45.29,
45.381

In order to compare distributions, $-2\ln L$, AIC (Akaike Information Criterion), AICC (Akaike Information Criterion Corrected), BIC (Bayesian Information Criterion), K-S Statistics (Kolmogorov-Smirnov Statistics) for two real data sets have been computed. The formulae for computing AIC, AICC, BIC, and K-S Statistics are as follows:

$$AIC = -2\ln L + 2k\,,\quad AICC = AIC + \frac{2k(k+1)}{(n-k-1)},$$

$$BIC = -2\ln L + k\ln n \text{ and}$$

$D = \underset{x}{\operatorname{Sup}}\left|F_n(x) - F_0(x)\right|$, where k = the number of parameters, n = the sample size and $F_n(x)$ is the empirical distribution function.

The best distribution corresponds to lower $-2\ln L$, AIC, AICC, BIC, and K-S statistics.

Table 1. MLE's, - 2ln L, AIC, AICC, BIC, and K-S Statistics of the fitted distributions of data sets 1 and 2

	Model	Parameter estimate	$-2\ln L$	AIC	AICC	BIC	K-S statistic
	Akash	1.1569	59.5	61.7	61.7	62.5	0.3205
Data 1	Lindley	0.8161	60.5	62.5	62.7	63.5	0.3410
	Exponential	0.5263	65.7	67.7	67.9	68.7	0.3895
	Akash	0.09706	240.7	242.7	242.8	244.1	0.2664
Data 2	Lindley	0.06299	254.0	256.0	256.1	257.4	0.3332
	Exponential	0.03246	274.5	276.7	276.7	277.9	0.4264

It can be easily seen from above table that the Akash distribution is better than the Lindley and exponential distributions for modeling life time data and thus Akash distribution should be preferred to exponential distribution and Lindley distributions for modeling lifetime data-sets.

13. CONCLUSIONS

A one parameter lifetime distribution named, "Akash distribution" has been proposed. Its mathematical properties including shape, moments, skewness, kurtosis, hazard rate function, mean residual life function, stochastic ordering, mean deviations, order statistics have been discussed. The condition under which Akash distribution is over-dispersed, equi-dispersed, and under-dispersed are presented along with the conditions under which exponential and Lindley distributions are over-dispersed, equi-dispersed and under-dispersed. Further, expressions for Bonferroni and Lorenz curves, Renyi entropy measure and stress-strength reliability of the proposed distribution have been derived. The method of moments and the method of maximum likelihood estimation have also been discussed for estimating its parameter. Finally, the goodness of fit test using K-S Statistics (Kolmogorov-Smirnov Statistics) for two real lifetime data- sets have been presented to illustrate its applicability and superiority over exponential and Lindley distributions.

ACKNOWLEDGEMENTS

The author is thankful to the anonymous reviewer for some useful comments.

REFERENCES

1. Abouammoh, A.M., Alshangiti, A.M. and Ragab, I.E. (2015): A new generalized Lindley distribution, *Journal of Statistical Computation and Simulation,* preprinthttp://dx.doi.org/10.1080/00949655.2014.995101.
2. Alkarni, S. (2015): Extended Power Lindley distribution-A new Statistical model for non- monotone survival data, *European journal of statistics and probability*, 3(3), 19 – 34.
3. Ashour, S. and Eltehiwy, M. (2014): Exponentiated Power Lindley distribution, *Journal of Advanced Research,* preprint http://dx.doi.org/10.1016/ j.jare. 2014.08.005.
4. Bakouch, H.S., Al-Zaharani, B. Al-Shomrani, A., Marchi, V. and Louzad, F. (2012): An extended Lindley distribution, *Journal of the Korean Statistical Society*, 41, 75 – 85.
5. Bonferroni, C.E. (1930): Elementi di Statistca generale, Seeber, Firenze.
6. Deniz, E. and Ojeda, E. (2011): The discrete Lindley distribution-Properties and Applications, *Journal of Statistical Computation and Simulation*, 81, 1405 – 1416.
7. Elbatal, I., Merovi, F. and Elgarhy, M. (2013): A new generalized Lindley distribution, *Mathematical theory and Modeling,* 3 (13), 30-47.
8. Fuller, E.J., Frieman, S., Quinn, J., Quinn, G., and Carter, W. (1994): Fracture mechanics approach to the design of glass aircraft windows: A case study, SPIE Proc 2286, 419-430.
9. Ghitany, M.E., Atieh, B. and Nadarajah, S. (2008): Lidley distribution and its Application, *Mathematics Computing and Simulation*, 78, 493 – 506.

10. Ghitany, M., Al-Mutairi, D., Balakrishnan, N. and Al-Enezi, I. (2013): Power Lindley distribution and associated inference, *Computational Statistics and Data Analysis*, 64, 20 – 33.

11. Gross, A.J. and Clark, V.A. (1975): Survival Distributions: Reliability Applications in the Biometrical Sciences, John Wiley, New York.

12. Hussain, E. (2006): The non-linear functions of Order Statistics and Their Properties in selected probability models, Ph.D thesis, Department of Statistics, University of Karachi, Pakistan.

13. Lindley, D.V. (1958): Fiducial distributions and Bayes' theorem, *Journal of the Royal Statistical Society, Series B*, 20, 102 - 107.

14. Liyanage, G.W. and Pararai. M. (2014): A generalized Power Lindley distribution with applications, *Asian journal of Mathematics and Applications*, 1 – 23.

15. Merovci, F. (2013): Transmuted Lindley distribution, *International Journal of Open Problems in Computer Science and Mathematics*, 6, 63 – 72.

16. Nadarajah, S., Bakouch, H.S. and Tahmasbi, R. (2011): A generalized Lindley distribution, *Sankhya Series* B, 73, 331 – 359.

17. Oluyede, B.O. and Yang, T. (2014): A new class of generalized Lindley distribution with applications, *Journal of Statistical Computation and Simulation*, 85 (10), 2072 – 2100.

18. Pararai, M., Liyanage, G.W. and Oluyede, B.O. (2015): A new class of generalized Power Lindley distribution with applications to lifetime data, *Theoretical Mathematics &Applications,* 5(1), 53 – 96.

19. Renyi, A. (1961): On measures of entropy and information, in proceedings of the 4th berkeley symposium on Mathematical Statistics and Probability, 1, 547 – 561, Berkeley, university of California press.

20. Shaked, M. and Shanthikumar, J.G. (1994): Stochastic Orders and Their Applications, Academic Press, New York.

21. Shanker, R. and Mishra, A. (2013 a): A two-parameter Lindley distribution, *Statistics in Transition-new series*, 14 (1), 45- 56.

22. Shanker, R. and Mishra, A. (2013 b): A quasi Lindley distribution, *African Journal of Mathematics and Computer Science Research*, 6(4), 64 – 71.

23. Shanker, R., Sharma, S. and Shanker, R. (2013): A two-parameter Lindley distribution for modeling waiting and survival times data, *Applied Mathematics*, 4, 363 – 368.

24. Shanker, R., Hagos, F, and Sujatha, S. (2015): On modeling of Lifetimes data using exponential and Lindley distributions, *Biometrics & Biostatistics International Journal*, 2 (5), 1-9.

25. Sharma, V., Singh, S., Singh, U. and Agiwal, V. (2015): The inverse Lindley distribution- A stress-strength reliability model with applications to head and neck cancer data, *Journal of Industrial &Production Engineering*, 32 (3), 162 – 173.

26. Singh, S.K., Singh, U. and Sharma, V.K. (2014): The Truncated Lindley distribution- inference and Application, *Journal of Statistics Applications& Probability*, 3(2), 219 – 228.

27. Zakerzadeh, H. and Dolati, A. (2009): Generalized Lindley distribution, *Journal of Mathematical extension*, 3 (2), 13 – 25.

Chapter 4

Fluctuation-Model-Based Discrete Probability Estimation for Small Samples

Takashi Isozaki

Sony Computer Science Laboratories, Inc., Tokyo, Japan

ABSTRACT

A robust method is proposed for estimating discrete probability functions for small samples. The proposed approach introduces and minimizes a parameterized objective function that is analogous to free energy functions in statistical physics. A key feature of the method is a model of the parameter that controls the trade-off between likelihood and robustness in response to the degree of fluctuation. The method thus does not require the value of the parameter to be manually selected. It is proved that the estimator approaches the maximum likelihood estimator at the asymptotic limit. The effectiveness of the method in terms of robustness is demonstrated by experimental studies on point estimation for probability distributions with various entropies.

Keywords: Probability Estimation, Fluctuation Models, Helmholtz Free Energy, Canonical Distributions

1. INTRODUCTION

For categorical observational data analysis, it is often necessary to deal with multivariate systems since variables of such data generally depend on each other. Highly predictive statistical inference requires parameters that achieve low-entropy, so it is preferable to use data of many variables because the following relationship in Shannon entropies H of random variables X and Y holds if X and Y depend on each other: $H(X|Y) \le H(X)$ [1] , where $H(X)$ denotes Shannon entropy of X and $H(X|Y)$ denotes the conditional entropy of X given Y. These are respectively defined with marginal, joint, and conditional probability mass functions P as follows [1] :

$$H(X) = -\sum_i P(X=i)\log P(X=i),$$

$$H(X|Y) = -\sum_{i,j} P(X=i, Y=j)\log P(X=i|Y=j),$$

where i and j are indices of discrete states of X and Y. When we estimate probabilities in discrete probabilistic models with many-variables (e.g., Markov network and Bayesian network models [2]), statistical estimation of conditional and joint probabilities often needs exponentially large data because of the combinatorial explosion of binding events in variables. Therefore, the models are often inferred from insufficient data. The maximum likelihood (ML) method provides an estimated probability function, which of X with k discrete states is expressed by

$$P(X=i) = \frac{n_i}{n}, \quad i = 1, \cdots, k$$

$$\sum_{i=1}^{k} P(X=i) = 1,$$

where n and n_i respectively denote a sample size and the frequency of occurrences in i state. The estimated probability is correct in the large sample limit. However, the ML methods suffer from short size data, and few robust methods have been investigated for such data in estimation of discrete probability functions as far as we know, although many robust methods for outliers such as M-estimators have been developed [3] [4] in para- metric continuous distributions. The maximum entropy method [5] , which may be applied to small datasets, was originally appropriate for data with missing information.

In the present study, a new robust method is proposed for estimating discrete probability functions for small samples. The method uses a parameterized objective function based on Kullback-Leibler divergence [6] and Shannon entropy. The function has a similar form to the Helmholtz free energy function that appears in statistical physics [7] [8] . A key feature of the method is a model of the parameter that controls the trade-off between likelihood and robustness in response to the degree of data fluctuation. The method thus does not require the value of the parameter to be manually selected. This model is a modification of a preceding work [9] , in which the parameter is represented by an artificial model containing a free hyperparameter.

In the domain of machine learning, although several methods slightly similar to ours have been proposed [10] - [14] , there is a critical distinction between these methods and ours. Many studies that have applied free energy to statistical inference have not included the similar trade-off parameter or have treated it as a fixed value, a manually controlled parameter, or a free parameter. Regarding the existing methods, we thus consider that the potentials of free-energy-like functions have not been well extracted. Other similar methods have been developed in the context of robust estimation for outliers, in which a free parameter is introduced in an analogous fashion [3] [15] [16] . However, the problem of how to determine the value of the free parameters remains.

This paper is organized as follows. In the next section, an objective function with parameter β is introduced for robust estimation, and then probability functions obtained by the proposed method are shown to be formally equivalent to the canonical distributions that appear in statistical physics. In Section 3, a new representation of β is presented as a data-fluctuation-model, and the preferable asymptotic property of β is proved. In Section 4, some characteristic properties among quantities used in the proposed method are provided. In Section 5, we perform experiments using the proposed probability estimation method. In Section 6, conclusions regarding the estimation method are given.

2. PROBABILITY ESTIMATION WITH PARAMETER Β

Note that in this paper a capital letter (such as X) denotes a random discrete variable, a non-capital letter (such as x) denotes the special state of that variable, a bold capital letter (such as $\boldsymbol{Y}$) denotes a set of variables, and a bold non-capital letter (such as $\boldsymbol{y}$) denotes configurations of that set.

To construct a method for estimating finite-discrete-probability distributions of random variable X from sample set of finite size n, the following quantities are defined. $P(X)$ denotes a discrete-probability function estimated by a proposed method that is described below. A function of X is defined as $U_0(X)$ on the basis of the Kullback--Leibler (KL) divergence [6] between empirical functions and $P(X)$ as follows:

$$U_0(X) = D\left(P(X) \| \tilde{P}(X)\right) = \sum_x P(x) \log \frac{P(x)}{\tilde{P}(x)}, \tag{1}$$

where $\tilde{P}(X)$ is a empirical distribution function. For non-parametric discrete distributions, the empirical distributions are equivalent to relative frequencies; i.e., they are equivalent to the maximum likelihood (ML) dis- tributions. $\tilde{P}(X)$ can thus be replaced by ML distributions denoted by $\tilde{P}(X)$. An objective function $F(X)$ is defined as follows:

$$F(X) = U_0(X) - \frac{1}{\beta_0(X)} H(X), \tag{2}$$

where $U_0(X)$ is defined by Equation (1), $H(X)$ is the Shannon entropy [1] of the estimated functions given as $H(X) = -\sum_x P(x) \log P(x)$, and $\beta_0(X)$ is a parameter that is defined so that $0 < \beta_0(X) < \infty$. $U(X)$ and $\beta(X)$ are introduced for later convenience. $U(X)$ is represented by a cross entropy $H\left(P(X), \hat{P}(X)\right)$ and $\beta(X)$ is a normalization parameter of $\beta_0(X)$ as follow:

$$U(X) = H\left(P(X), \hat{P}(X)\right) = -\sum_x P(x) \log \hat{P}(x) \tag{3}$$

and

$$\beta(X) = \beta_0(X) / \left(1 + \beta_0(X)\right), \tag{4}$$

where $0 < \beta(X) < 1$. $F(X)$ can be rewritten by using $U(X)$ and $\beta(X)$ as:

$$F(X) = U(X) - \frac{1}{\beta(X)} H(X). \tag{5}$$

In addition, the following quantity $\epsilon(x)$ is defined as

$$\epsilon(x) = -\log \hat{P}(x). \tag{6}$$

$U(X)$ is also written with ϵ as $U(X)=\langle\epsilon\rangle(X)$; that is, $\langle\rangle$ denotes an expectation value in respect to $P(X)$.

The estimator of probability functions, $P(X)$, is defined so as to minimize Lagrangian L consisting of $F(X)$ as $\partial L/\partial P(X)=0$. L is expressed as

$$L=F+\lambda\left(\sum_x P(x)-1\right)=\frac{1}{\beta}\sum_x P(x)\log P(x)-\sum_x P(x)\log\hat{P}(x)+\lambda\left(\sum_x P(x)-1\right), \tag{7}$$

where λ is the Lagrange multiplier. $P(X)$ is thereby obtained as

$$P(x)=\frac{\exp\left(-\beta\left(-\log\hat{P}(x)\right)\right)}{\sum_{x'}\exp\left(-\beta\left(-\log\hat{P}(x')\right)\right)}=\frac{\exp\left(-\beta\epsilon(x)\right)}{\sum_{x'}\exp\left(-\beta\epsilon(x')\right)}, \tag{8}$$

where $\epsilon(x)$ as expressed in Equation (6) is used. Equation (8) is equivalent to a form known as the canonical distribution, which is also called Gibbs distribution, in statistical physics. The following equivalent form is more convenient for practical use:

$$P(x)=\frac{\left[\hat{P}(x)\right]^{\beta}}{\sum_{x'}\left[\hat{P}(x')\right]^{\beta}}. \tag{9}$$

For estimating conditional and joint-probability functions, conditional entropy $H(X|\mathbf{Y})$, which is defined as: $H(X|\mathbf{Y})=-\sum_{x,y}P(x,y)\log P(x|y)$, and conditional KL divergence:

$$D\left(P(X|\mathbf{Y})\|Q(X|\mathbf{Y})\right)=\sum_{x,y}P(x,y)\log\left(P(x|y)/Q(x|y)\right)$$

are used. β for $P(x|y)$ is defined as

$$\beta(X|\mathbf{Y})=\beta_0(X|\mathbf{Y})/\left(\beta_0(X|\mathbf{Y})+1\right).$$

The formula for estimating conditional probabilities is therefore obtained by using the conditional entropy and KL divergence and $\beta(X|\mathbf{Y})$ in the following form:

$$P(x|y) = \frac{\exp\left(-\beta\left(-\log \hat{P}(x|y)\right)\right)}{\sum_{x'} \exp\left(-\beta\left(-\log \hat{P}(x'|y)\right)\right)}. \quad (10)$$

Joint probability can be calculated by using Equations (8) and (10) and the definite relation $P(X,Y) = P(X|Y)P(Y)$. In general, it is calculated using decomposition rules such that

$$P(X_1, X_2, \cdots, X_n) = P(X_n | X_{n-1}, \cdots, X_2, X_1) \cdots P(X_2 | X_1) P(X_1).$$

3. MODEL OF Β

$P(X)$ approaches $\hat{P}(X)$ when β in Equation (9) approaches 1. On the other hand, $P(X)$ approaches the uniform distribution if $\hat{P}(X) > 0$ and β approaches 0. If β close to 1 represents that the data size is suffi- ciently large and if β close to 0 represents that the data size is very small, β has favorable properties for accurate and robust estimation. This is because the ML estimators generally have preferable consistency and asymptotic efficiency and the distributions close to the uniform can be regarded as the ones that have robustness for small size data. Before β is defined, the following quantity $P_n^G(X)$ with n data size is defined by a geometric mean as

$$P_n^G(X) = Z_n^G \left(\prod_{i=0}^{n} P_i(X) \right)^{\frac{1}{n+1}}, \quad (11)$$

where Z_n^G denotes a normalization constant, and $P_i(X)$ denotes the estimated function obtained from Equation (9) with initial i data. It is defined that $P_0^G(X) = P_0(X) = 1/|X|$, where $|X|$ denotes the number of states of variable X. Fluctuation δ of X with n data is given as

$$\delta_n(X) = \log P_{n-1}^G(X) - \log \hat{P}(X).$$

Then, $\beta_0(X)$ for n data is defined by using an expectation value of $\delta_n(X)$ with respect to $P_{n-1}^G(X)$, i.e., the following KL divergence:

$$\beta_0(X) = 1 \Big/ D\left(P_{n-1}^G(X) \| \hat{P}(X)\right) = 1 \Big/ \sum_x P_{n-1}^G(x) \log \frac{P_{n-1}^G(x)}{\hat{P}(x)}, \tag{12}$$

where $n \geq 1$, and $\hat{P}(X)$ is the ML estimator function obtained from n data. It is assumed that

$D\left(P_{n-1}^G(X) \| \hat{P}(X)\right) > 0$. The normalized β_0, that is, β, is defined by Equation (4). Note that the canonical distribution $P(x)$ expressed by Equation (9) can be determined, without any free parameters, by using Equations (9), (11), (12), (4), and $P(X) = 1/|X|$ for data size $n = 0$. It is also defined that in Equation (10) for conditional data size $n = 0$ given $\boldsymbol{Y} = \boldsymbol{y}$, $P(x|\boldsymbol{y}) = 1/|X|$ for any $\boldsymbol{y}$ in the same manner as $P(x)$.

Objective function F is rewritten in the same form as that in statistical mechanics as follows:

$$F = U - \frac{1}{\beta} H = -\frac{1}{\beta} \log Z, \tag{13}$$

where Z is the partition function, which is a similar function well known in statistical mechanics, defined for single or multivariate probabilities as

$$Z(\boldsymbol{X}) = \sum_x \left[\hat{P}(\boldsymbol{x})\right]^\beta, \tag{14}$$

and for conditional probabilities as

$$Z(X|\boldsymbol{Y}) = \sum_x \left[\hat{P}(x|\boldsymbol{Y})\right]^\beta. \tag{15}$$

A significant feature of β is confirmed as follows. The lemma needed for this proof is stated as

Lemma 1.

$P_i(X)$ is denoted as the canonical distribution estimated from Equation (8) with i data. For data size $n \to \infty$, $P_n^G(x)$, defined by Equation (11), converges to a definite value $P^G(x)$ when $P_i(x) > 0$ for inte- gers i such that $i \geq 0$ and any state x of X.

Proof.

$$\begin{aligned}&\log P_n^G(x)-\log P_{n-1}^G(x)\\&=\frac{1}{n+1}\sum_{i=0}^{n}\log P_i(x)-\frac{1}{n}\sum_{i=0}^{n-1}\log P_i(x)+\log Z_n^G-\log Z_{n-1}^G\\&=\left(\frac{n}{n+1}-1\right)\log P_{n-1}^G+\frac{1}{n+1}\log P_n(x)+\log Z_n^G-\log Z_{n-1}^G.\end{aligned} \quad (16)$$

Because $P_{n-1}^G(x)>0$ and $P_n(x)>0$, $\log P_{n-1}^G(x)$ and $\log P_n(x)$ are definite values. Thus both terms on the right-hand side of Equation (16) converge to $\log Z_n^G-\log Z_{n-1}^G$, which is a constant if $n\to\infty$. Hence, $P_n^G(x)=bP_{n-1}^G$, with constant b, can be written. Therefore, $b=1$ and $P_n^G(x)\to P^G(x)$ in order that $P_n^G(x)$ converges not to 0 or 1 but to definite values. □

Theorem 1. At the asymptotic limit (i.e., large sample limit), β converges to 1 when $P_i(x)>0$ for integers i such that $i\geq 0$ and any state x, where $P_i(x)$ is denoted as the canonical distribution represented by Equation (8) from i data.

Proof. According to Lemma 1, $P_n^G(x)\to P^G(x)$ at the limit $n\to\infty$, where $P^G(x)$ is a definite value for any x. The following Equation is thus obtained from Equation (11) with $P_i(x)$ as follows:

$$\log P^G(x)=\lim_{n\to\infty}\frac{1}{n+1}\sum_{i=0}^{n}\log P_i(x)+\text{constant}, \quad (17)$$

$P_n(x)$ thus converges to a definite value, and $P^G(x)$ converges to $P_n(x)$ (and the constant in Equation (17) goes to 0) at $n\to\infty$. Meanwhile, ML estimator $\hat{P}(x)$ converges to true distribution $P_t(x)$ due to the consistency of the ML estimators, and β thus converges to a definite value according to Equation (9). $P_n(x)$ at $n\to\infty$ is denoted as $P_\infty(x)$. Therefore, the following Equation is derived from Equations (4), (9) and (12) at $n\to\infty$,

$$\frac{1}{\beta_0}=\frac{1-\beta}{\beta}=D\left(P_\infty(X)\|P_t(X)\right)=\sum_x\frac{\left[P_t(x)\right]^\beta}{Z}\log\frac{\left[P_t(x)\right]^{\beta-1}}{Z}=\text{a definite value} \quad (18)$$

for $P(x)>0$ and any state x. Equation (18) requires $\beta \to 0$ or $\beta \to 1$ in order that $[P_t(x)]^{\beta}$ or $[P_t(x)]^{\beta-1}$ is a constant for any probability distribution $P_t(x)$. However, $\beta \to 0$ does not satisfy Equation

(18), while $\beta \to 1$ satisfies it. Accordingly, $\beta \to 1$ at the asymptotic limit. □

According to Theorem 1, the more data are obtained, the more β approaches 1, and the more the estimator approaches the ML estimator. The estimator that is obtained by the proposed method therefore has the same preferable asymptotic properties, namely, consistency and efficiency, as the ML estimators have. For insuffi- cient data size, β_0 is probably small due to the influence of the uniform distributions given by Equation (12), so β is also small. The estimated probability functions by the proposed method are thus interpreted as adap- tively tempered ML estimator functions in response to the degree of data fluctuation. The proposed estimation method does thereby not require manually selecting the value of parameter β, and is called "ATML," which is abbreviated as the "adaptively tempered ML" method. ATML has an advantage of simpleness over methods that need complicated algorithms (e.g., [17]).

The role of β can be seen as a trade-off parameter between likelihood and robustness by referring to another expression of Equation (2) as follows:

$$F = \beta D\left(P(X)\|\hat{P}(X)\right) + (1-\beta)D\left(P(X)\|P_u(X)\right) + \text{constant}, \quad (19)$$

where $P_u(X)$ denotes the uniform distribution function, which contributes to the robustness, while $\hat{P}(X)$ contributes to the likelihood. Additionally, objective function F can also be interpreted as a KL-based diver- gence measure since β is also represented by a KL divergence.

ATML has an analogy with statistical physics, since the canonical distribution for the estimator is obtained from Equation (8). Actually, U, H, β, and F respectively play similar roles to (internal) energy, entropy, (inverted) temperature, and Helmholtz free energy in statistical physics. Solving Equation $\partial L/\partial P(X)=0$ for obtaining $P(X)$ mathematically corresponds to employing the minimum-free-energy (MFE) principle [7] in thermal physics.

ATML may seem analogous to Jaynes' maximum entropy (ME) methods [5] , which are well known as least- biased inference methods. However, the constraints on which ME methods are based may not be reliable for small samples and thus may be biased, although this kind of bias is not usually considered. On the other hand, ATML is thus designed so that even the bias can be corrected by using parameter β.

4. CHARACTERISTIC PROPERTIES OF ATML

The canonical distribution expressed as Equation (8) can provide some characteristic properties of ATML, which are similar to those in statistical physics. The following notations are defined for later convenience. Pro- bability mass functions that are estimated by ATML, denoted by P_k, have discrete states denoted as index k. The corresponding ML estimator is denoted by $\hat{P}_k$, and $\{\beta,U\}$ is used instead of $\{\beta_0,U_0\}$.

In statistical physics, (inverted) temperature β is usually defined as [7] [8]

$$\beta := \frac{\partial H}{\partial U}. \tag{20}$$

If the canonical distribution of P_k, which takes the form of Equation (8), is used, Equation (20) is auto- matically satisfied as follows:

Lemma 2. $H = \beta U + \log Z$ under the MFE condition, where H, U, β, and Z are defined in the previous sections.

Proof. Since probability mass function P_k has a canonical form under the MFE condition, it follows that

$$\begin{aligned} H &= -\sum_k P_k \log P_k = -\sum_k \frac{\hat{P}_k^\beta}{Z} \log \frac{\hat{P}_k^\beta}{Z} \\ &= \beta \cdot \left(-\sum_k \frac{\hat{P}_k^\beta}{Z} \log \hat{P}_k \right) + \left(\sum_k \frac{\hat{P}_k^\beta}{Z} \right) \cdot \log Z = \beta U + \log Z. \end{aligned} \tag{21}$$

Theorem 2. Equation (20) is automatically satisfied under the MFE condition.

Proof. Partially differentiating both sides of Equation (21) with respect to U gives

$$\frac{\partial H}{\partial U} = \beta + U \frac{\partial \beta}{\partial U} + \frac{\partial}{\partial U} \log Z. \tag{22}$$

$$\frac{\partial}{\partial U}\log Z = \frac{\partial \beta}{\partial U}\frac{\partial}{\partial \beta}\log Z. \tag{23}$$

$$\frac{\partial}{\partial \beta}\log Z = \frac{1}{Z}\frac{\partial}{\partial \beta}\left(\sum_k \hat{P}_k^{\beta}\right) = \frac{1}{Z}\sum_k \hat{P}_k^{\beta}\log \hat{P}_k = \sum_k \frac{\hat{P}_k^{\beta}}{Z}\log \hat{P}_k = -U.$$

It follows that

$$\frac{\partial H}{\partial U} = \beta. \quad \square$$

In the same way, it can be proved that $\beta_0 = \partial H / \partial U_0$.

$\langle \epsilon^2 \rangle - \langle \epsilon \rangle^2$, which is called energy fluctuations in statistical mechanics, is shown to have the following relation, where $\langle \rangle$ denotes an expectation value with respect to the canonical distributions.

$$\langle \epsilon^2 \rangle - \langle \epsilon \rangle^2 = -\frac{\partial U}{\partial \beta}. \tag{24}$$

In regard to ϵ defined in the proposed estimation method, namely, Equation (6), the same relation as that shown here is satisfied as follows:

$$\begin{aligned}
-\frac{\partial U}{\partial \beta} &= \sum_k \left(\log \hat{P}_k\right)\left\{\frac{1}{Z}\hat{P}_k^{\beta}\log \hat{P}_k - \left(\frac{1}{Z^2}\right)\sum_m \frac{\partial \hat{P}_m^{\beta}}{\partial \beta}\right\} \\
&= \sum_k \left(\log \hat{P}_k\right)\left\{\frac{\hat{P}_k^{\beta}}{Z}\log \hat{P}_k - \frac{\hat{P}_k^{\beta}}{Z}\sum_m \frac{\hat{P}_m^{\beta}}{Z}\log \hat{P}_m\right\} \\
&= \sum_k \frac{\hat{P}_k^{\beta}}{Z}\log \hat{P}_k\left(\log \hat{P}_k - \sum_m \frac{\hat{P}_m^{\beta}}{Z}\log \hat{P}_m\right) \\
&= \langle \epsilon^2 \rangle - \langle \epsilon \rangle^2,
\end{aligned}$$

Equation (24) is therefore proved.

Fisher information $\tilde{I}(\beta)$ with a parameter β is defined in the usual way as

$$\tilde{I}(\beta) := \sum_k f_k(\beta)\left(\frac{\partial}{\partial \beta}\log f_k(\beta)\right)^2, \tag{25}$$

where f is the likelihood function. We define tempered Fisher information $I(\beta)$ as Fisher information where the likelihood function is replaced with the canonical distributions with parameter β. It is shown that $I(\beta)=\langle\epsilon^2\rangle-\langle\epsilon\rangle^2$ as follows:

$$I(\beta)=\sum_k \frac{\hat{P}_k^\beta}{Z}\left(\frac{\partial}{\partial\beta}\log\frac{\hat{P}_k^\beta}{Z}\right)^2=\sum_k \frac{\hat{P}_k^\beta}{Z}\left(\log\hat{P}_k-\frac{1}{Z}\sum_m \hat{P}_m^\beta\log\hat{P}_m\right)^2$$
$$=\sum_k \frac{\hat{P}_k^\beta}{Z}\left(-\log\hat{P}_k\right)^2-\left(\sum_k \frac{\hat{P}_k^\beta}{Z}\left(-\log\hat{P}_k\right)\right)^2=\langle\epsilon^2\rangle-\langle\epsilon\rangle^2. \tag{26}$$

The tempered Fisher information is therefore identical to Equation (24).

It is noteworthy that ATML has other mathematical similarities with statistical physics. That is, the same relationships that appear in statistical physics listed as follows hold.

- The following relation is easily derived from the definition of partition function Z:

$$U=-\frac{\partial}{\partial\beta}\log Z. \tag{27}$$

- The following relation, known as the Gibbs-Helmholtz relation, is derived from Equations (13) and (27) as

$$U=\frac{\partial}{\partial\beta}(\beta F). \tag{28}$$

- The following relation is simply obtained from Equations (5) and (28) as

$$H=\beta^2\frac{\partial F}{\partial\beta}. \tag{29}$$

- The tempered Fisher information is represented by the second-order differential of the partition function for β as

$$I(\beta)=\langle\epsilon^2\rangle-\langle\epsilon\rangle^2=\frac{\partial^2}{\partial\beta^2}\log Z. \tag{30}$$

5. EXAMPLES

Numerical experiments are performed to demonstrate the robustness of ATML for small samples, in comparison with the ML and ME methods [5] .

X is assumed to have three internal states and four probability mass functions with a variety of entropies denoted as $H(X)$ in natural logarithms as

1. $P(x=0)=0.431, P(x=1)=0.337, P(x=2)=0.232, H(X)=1.07$,
2. $P(x=0)=0.677, P(x=1)=0.206, P(x=2)=0.117, H(X)=0.841$,
3. $P(x=0)=0.851, P(x=1)=0.117, P(x=2)=0.0320, H(X)=0.498$,
4. $P(x=0)=0.9898, P(x=1)=0.00810, P(x=2)=0.00210, H(X)=0.0621$.

Data from each function was sampled, and probabilities were estimated from given data sets with various data sizes. According to convention, averaged outputs $\langle X\rangle$ were set as the constraint in the ME method as follows: $\langle X\rangle := (1/N)\sum_{d=1}^{N} X_d$, where X_d denotes d-th sample's output, and N denotes sample size. After that, true and

estimated probabilities were compared by using KL divergence as a metric with the following form:

$$D(P(X)\|P_e(X)) = \sum_x P(x)\log\frac{P(x)}{P_e(x)}, \tag{31}$$

where $P(X)$ is the true distribution, and $P_e(X)$ is the distribution estimated by ML, ME, or ATML. For avoiding zero probabilities, probabilities were smoothed by adding 0.0001 to the counts.

The KL divergences are shown in Figure 1, where they are averaged values from 100 samples at each sample size from identical distributions. It can be seen that the ML estimators are inferior to ATML due to overfitting, except for the distribution having very small entropy. Even the degree of superiority of the ML estimation in (d) is relatively smaller than that of inferiority in other distributions. The ME estimators showed the opposite be- haviors to those by the ML estimators, and showed some relatively poor results in large-sample regions. ML methods tend to fit data and can thus more accurately estimate distributions with very low entropies than others in small-sample cases. For example, if a true entropy equals zero, the ML method can estimate the exact true distribution from only one sample. On the other hand, ME methods tend to increase entropies and can thereby accurately estimate distributions with high entropies close to the uniform distributions. Hence, the ML method tends to overfit data, and the ME method tends to underfit data in the view of misestimation. Even so, ATML showed relative stability in terms of both sample sizes and distributions. This result indicates the effectiveness of ATML as a probability estimation method.

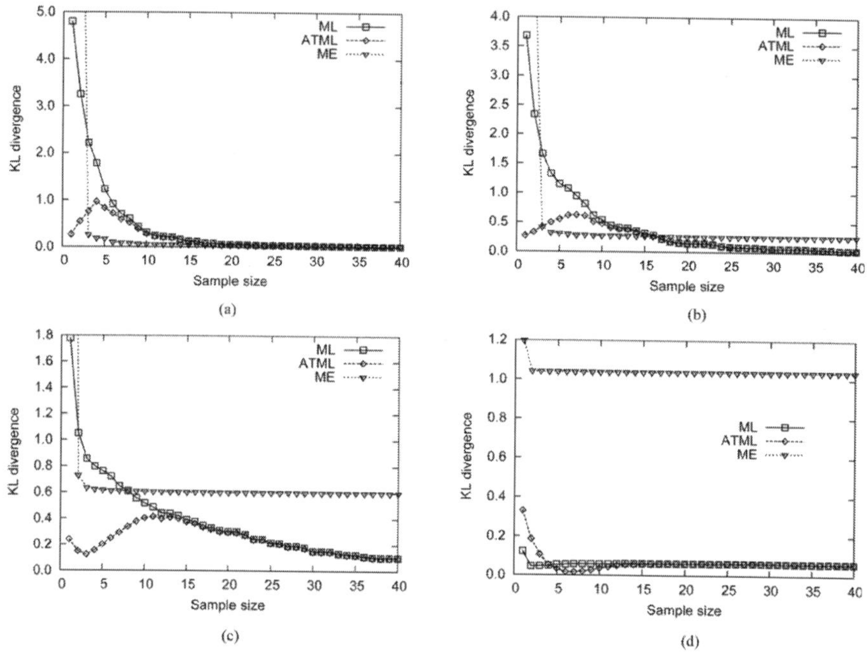

Figure 1. KL divergences between true probability mass functions and probability mass functions estimated by using ML, ATML, and ME. The horizontal axes denote sample sizes. H denotes Shannon entropy in natural logarithms. (a) H = 1:07; (b) H = 0:841; (c) H = 0:498; (d) H = 0:0621.

6. CONCLUSION

A robust method for estimating discrete probability functions, called "adaptively tempered maximum likelihood" method (ATML for short), is proposed. The estimators obtained in this method minimize a parameterized objective function similar to Helmholtz free energies that appear in statistical physics. The key feature of the proposed method is a model of the parameter as a fluctuation of finite size data. The parameter that is modeled plays an important role in determining the appropriate trade-off between likelihood and robustness in response to the degree of the fluctuations. ATML does thereby not require manually selecting the value of the parameter. It is also proved that the obtained estimator approaches the maximum likelihood estimator at the asymptotic limit. The effectiveness of ATML in terms of robustness was demonstrated by experimental studies on point estimation for probability distributions with various entropies.

ACKNOWLEDGEMENTS

The author thanks Mario Tokoro and Hiroaki Kitano of Sony Computer Science Laboratories, Inc. for their support as well as Te Sun Han, Hiroshi Nagaoka, Tomohiro Ogawa and Jun Suzuki of the University of Electro- Communications for their valuable comments and discussions.

REFERENCES

1. Shannon, C.E. (1948) A Mathematical Theory of Communication. Bell Systems Technical Journal, 27, 379-423, 623-656.
2. Pearl, J. (1988) Probabilistic Reasoning in Intelligent Systems. Morgan Kaufmann, San Mateo, CA.
3. Basu, A., Harris, I.R., Hjort, N.L. and Jones, M.C. (1998) Robust and Efficient Estimation by Minimising a Density Power Divergence. Biometrika, 85, 549-559.
4. Beran, R. (1977) Minimum Hellinger Distance Estimates for Parametric Models. Annals of Statistics, 5, 445-463.
5. Jaynes, E.T. (1957) Information Theory and Statistical Mechanics. Physical Review, 106, 620-630.
6. Kullback, S. and Leibler, R.A. (1951) On Information and Sufficiency. Annals of Mathematical Statistics, 22, 79-86.
7. Callen, H.B. (1985) Thermodynamics and an Introduction to Thermostatistics. 2nd Edition, John Wiley & Sons, Hoboken, NJ.
8. Kittel, C. and Kroemer, H. (1980) Thermal Physics. W. H. Freeman, San Francisco, CA.
9. Isozaki, T., Kato, N. and Ueno, M. (2009) "Data Temperature" in Minimum Free Energies for Parameter Learning of Bayesian Networks. International Journal on Artificial Intelligence Tools, 18, 653-671.
10. Hofmann, T. (1999) Probabilistic Latent Semantic Analysis. Proceedings of Conference on Uncertainty in Artificial Intelligence (UAI-99), Stockholm, 30 July-1 August 1999, 289-296.
11. LeCun, Y. and Huang, F.J. (2005) Loss Functions for Discriminative Training of Energy-Based Models. Proceedings of International

Workshop on Artificial Intelligence and Statistics (AISTATS-05), Barbados, 6-8 January 2005, 206-213.

12. Pereira, F., Tishby, N. and Lee, L. (1993) Distributional Clustering of English Words. In: Proceedings of Annual Meeting on Association for Computational Linguistics (ACL-93), Association for Computational Linguistics, Stroudsburg, 183-190.

13. Ueda, N. and Nakano, R. (1995) Deterministic Annealing Variant of the EM Algorithm. Proceedings of Advances in Neural Information Processing Systems 7 (NIPS 7), Denver, 29 November-1 December 1994, 545-552.

14. Watanabe, K., Shiga, M. and Watanabe, S. (2009) Upper Bound for Variational Free Energy of Bayesian Networks. Machine Learning, 75, 199-215.

15. Jones, M.C., Hjort, N.L., Harris, I.R. and Basu, A. (2001) A Comparison of Related Density-Based Minimum Divergence Estimators. Biometrika, 88, 865-873.

16. Windham, M.P. (1995) Robustifying Model Fitting. Journal of the Royal Statistical Society B, 57, 599-609.

17. Pöschel, T., Ebeling, W., Frömmel, C. and Ramírez, R. (2003) Correction Algorithm for Finite Sample Statistics. The European Physical Journal E, 12, 531-541.

Chapter 5

A Quantitative Comparative Study of Back Projection, Filtered Back Projection, Gradient and Bayesian Reconstruction Algorithms in Computed Tomography (CT)

Zoubeida Messali [1,2], Nabil Chetih [3], Amina Serir [4], Abdelwahab Boudjelal [2]

[1]*Department of Electronic, University of El Bachir El Ibrahimi, Bordj Bou Arreridj, Algeria*

[2]*Laboratoire de Génie Electrique (LGE), Université de M'Sila, M'Sila, Algérie*

[3]*Welding and NDT Research Center (CSC), Cheraga, Algiers, Algeria*

[4]*Laboratoire de Traitement d'Images et de Rayonnement (LTIR), Université des Sciences et de la Technologie Houari Boumediene, Alger, Algerie*

ABSTRACT

Images of the inside of the human body can be obtained noninvasively using tomographic acquisition and processing techniques. In particular, these techniques are commonly used to obtain X-ray images of the human body. The reconstructed images are obtained given a set of their projections, acquired using reconstruction techniques. A general overview of analytical and iterative methods of reconstruction in computed tomography (CT) is presented in this paper, with a special focus on Back Projection (BP), Filter Back Projection (FBP), Gradient and Bayesian maximum a posteriori (MAP) algorithms. Projections (parallel beam type) for the image reconstruction are calculated analytically by defining two phantoms: Shepp-Logan phantom head model and the standard medical image of abdomen with coverage angle ranging from 0 to ± 180° with rotational increment of 10°. The original images are grayscale images of size 128 128, 256 256, respectively. The simulated results are compared using quality measurement parameters for various test cases and conclusion is achieved. Through these simulated results, we have demonstrated that the Bayesian (MAP) approach provides the best image quality and appears to be efficient in terms of error reduction.

Keywords: Computed tomography, Bayesian approach, Reconstruction techniques, Filter Back Projection (FBP), Gradient algorithm

1. INTRODUCTION

During the last decade, there has been an avalanche of publications on different aspects of computed tomography [1-5]. The scientific word tomography means reconstruction from slices. It is an imaging technique which uses the absorption of X-rays by a number of organs in the body. This creates a shadow picture (projection). By sending the X-rays from different angles around the object, a collection of shadow pictures (projections) is created. Combining these pictures results in a tomogram, which is basically the image of a cross-section of the body. The impact of this technique in diagnostic medicine has been revolutionary, since it has enabled doctors to view internal organs with unprecedented precision and safety to the patient. A normal radiography shows mainly the bones and organs within the body. They absorb a major part of the X-rays. The absorption of the X-rays by the neighboring tissue is a lot smaller. This is the reason bones become visible on film, similar to a shadow picture.

Given the enormous success of X-ray computed tomography (CT), it is not surprising that in recent years much attention has been focused on extending this image formation technique to nuclear medicine [6] and magnetic resonance on the one hand; and ultrasound and microwaves on the other [7]. In nuclear medicine, our interest is in reconstructing a cross-sectional image of radioactive isotope distributions within the human body; and in imaging with magnetic resonance we wish to reconstruct the magnetic properties of the object. In both these areas, the problem is still the same and can be set up as reconstructing an image from its projections.

In tomography, images coming from the computed tomography scan are combined to create a visualization of the scanned body. To create this image we use reconstruction techniques. The most effective type of reconstruction technique is able to reconstruct good quality CT images even when the projected data are noisy. The conventional algorithms of image reconstruction for CT are Back Projection BP and Filtered Back Projection FBP reconstruction techniques which are analytical reconstruction methods. In Filtered Back Projection methodology, Fourier Slice Theorem is made into use for Image reconstruction [8]. Although for now the Filtered Back Projection algorithm is most widely used by manufacturers, efforts have been made to make iterative methods popular again due to their unique advantages, such as their performances with incomplete noisy data. The algebraic reconstruction technique (ART) [1-3], the simultaneous algebraic reconstruction technique (SART) [4] and the simultaneous iterative reconstruction technique (SIRT) [5] are a few of those iterative methods. The algebraic reconstruction technique (ART) [9] was also the first iterative algorithm to be used in CT. ART can be seen as a linear algebraic problem. It is based on the simple assumption that calculating the cross-section can be done by making algebraic equations for the array of unknowns referring to the measured projection data. As well known, simultaneous algebraic reconstruction technique (SART) was proposed as a major refinement of ART, also used in the X-ray tomography, which combines the positive aspect of SIRT and ART. One of the most important iterative reconstruction techniques in CT is the Gradient algorithm [6]. It is considered as an optimization method. More recently, stochastic methods, which are based on the Bayesian framework, have been successfully used in CT [10-13].

This article presents the basics of the more widely used algorithms: Back Projection (BP), Filtered Back Projection (FBP), Gradient algorithms and a special focus on a Bayesian maximum a posteriori (MAP) approach. It is not an exhaustive presentation of reconstruction algorithms. Therefore, this paper is aimed to establish a quantitative comparative study of analytical and iterative reconstruction techniques in order to reduce the number of iterations and enhance the image quality. Two Phantoms are used to test the results of the research: the Shepp-Logan Phantom and the standard medical image of abdomen. So, physical phenomena such as attenuation, scattering, and so forth are not under consideration here. The original images are gray scale images of size 128×128, 256×256 respectively.

After introducing the basic concepts and the supporting theory in section 2, we present the reconstruction techniques in sections 3 and 4. We focus on developing a Bayesian (MAP) approach, in which the reconstruction is computed by maximizing an associated objective function via an iterative algorithm. Following this, in section 5, we explain how to evaluate the quality of the reconstruction. Section 6 provides some test results and comparisons. Finally, in the last section our conclusions can be read.

2. BASIC FORMULATION

Fundamentally, tomographic imaging deals with reconstructing an image from its projections. In the strict sense of the word, projections are a set of measurements of the integrated values of some parameter of the object-integrations being along straight lines through the object and being referred to as line integrals. A line integral, as the name implies, represents the integral of some parameter of the image along a line as illustrated in Figure 1.

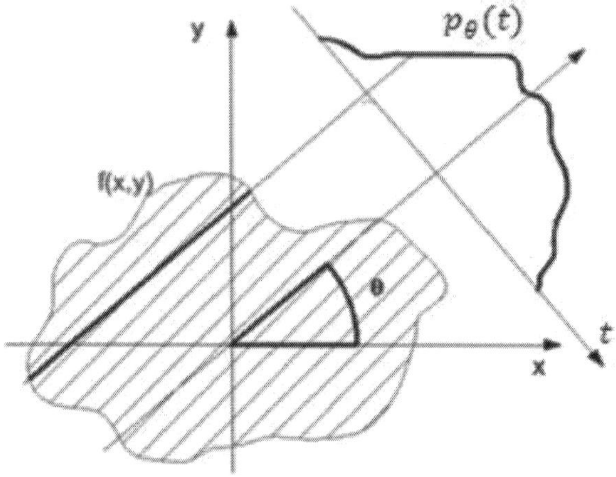

Figure 1. An object, f(x, y), and its projection, pθ(t), are shown for an angle of θ

The projections (or line integral), as shown in Figure 1, are defined as:

$$p_\theta(t) = \int_{(\theta,t)} f(x,y)ds \tag{1}$$

The function $p_\theta(t)$ is known as the Radon transform of the function $f(x,y)$, where $f(x,y)$ represents the 2D image to be reconstructed, and (θ,t) the parameters ofeach line integral. $p_\theta(t)$ is also known as a sinogram of the image. A sinogram as shown in Figure 2, is a 2D image, in which the horizontal axis represents the count location on the detector, and the vertical axis corresponds to the angular position of the detector.

Using a delta function, $p_\theta(t)$ can be rewritten as:

$$p_\theta(t) = \iint_{-\infty}^{+\infty} f(x,y)\delta(x \cos\theta + y \sin\theta - t)dx\, dy \tag{2}$$

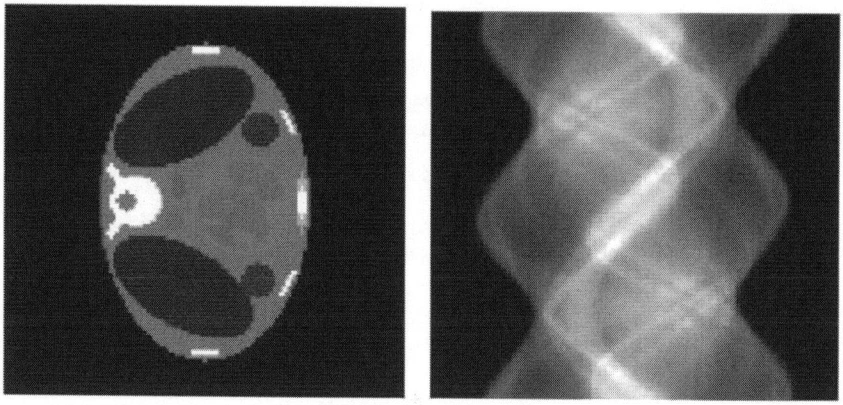

Figure 2. (Left) Shepp-Logan phantom head model (128 × 128 pixels). (Right) Corresponding sinogram, with coverage angle ranging from 0° to 180 with rotational increment of 10°

It is well known that, from knowledge of the sinogram $p_\theta(t)$, one can readily reconstruct the image $f(x,y)$ by use of computationally efficient and numerically stable algorithms. The algorithm that is currently being used in almost all applications of straight ray tomography is Filtered Back-Projection (FBP) algorithm. This algorithm will be described in the following.

3. ANALYTICAL RECONSTRUCTION METHODS

Analytical reconstruction methods consider continuoustomography. So the problem is: given the sinogram $p_\theta(t)$ we want to recover the object described in (x,y) coordinates. The most common types of image analytical reconstructions are Back Projection (BP) and Filtered Back Projection (FBP). For the sake of clarity, we first describe the backprojection (BP) algorithm.

$$b(x,y) = \hat{f}(x,y) = \int_0^\pi p_\theta(t)d\theta \tag{3}$$

3.1. Back-projection Algorithm (BP)

Backprojection represents the accumulation of the ray-sums of all the rays that pass through any point. Applying backprojection to projection data is called the summation algorithm. In ideal conditions (in particular, without attenuation), the projections acquired at angles between π radians (180°) and 2π radians (360°) do not provide new information, because they are only the symmetric of the projections acquired at angles between 0 and π. This is why the bounds of the integral are limited between 0 and π radians. The main point of this part is that the backprojection operation is not the inverse of the projection operation. This means that applying backprojection to the function $p_\theta(t)$ does not yield $f(x,y)$ but a blurred $f(x,y)$ as it will be shown in our simulation results.

3.2. Filtered Back-projection (FBP) Algorithm

It is possible to derive reconstruction algorithms from the so-called Fourier slice theorem [8]. This theorem relates the Fourier transform of a projection to the two-dimensional Fourier transform of the object which is to be reconstructed. Thus given the Fourier transform of a projection at enough angles the projections could be assembled into a complete estimate of the two dimensional transform and then simply inverted to arrive at an estimate of the object. The algorithm that is derived by using the Fourier Slice Theorem is the Filtered Back Projection algorithm. It has been shown to be extremely accurate and amenable to fast implementation. The FBP can be understood as follows: each line of the sinogram is filtered first in order to decrease the blur of the reconstructed image obtained by BP, and the filtered projections are backprojected. This is mathematically expressed as:

$$f(x,y) = \int_0^\pi \hat{p}_\theta(t)d\theta \tag{4}$$

Where $\hat{p}_\theta(t)$ is the filtered version of $p_\theta(t)$ with the ramp filter which gives a weight proportional to its frequency to each of the components. So, the relationship between $\hat{p}_\theta(t)$ and $p_\theta(t)$ is expressed as

$$\hat{p}_\theta(t) = \int_{-\infty}^{\infty} P_\theta(v) \cdot |v| \cdot e^{j2\pi vt} dv \tag{5}$$

Where $P_\theta(v)$ is 1D Fourier Transform of $p_\theta(t)$, multiplying with $|v|$ gives ramp filtering, integrating the expression from $-\infty$ to ∞ gives inverse 1D Fourier Transform. Substituting (5) in (4), yields:

$$f(x,y) = \int_0^\pi \left(\int_{-\infty}^{\infty} P_\theta(v) \cdot |v| \cdot e^{j2\pi vt} dv\right)d\theta \tag{6}$$

Therefore, the complete Filtered Back Projection algorithm can be viewed in Figure 3.

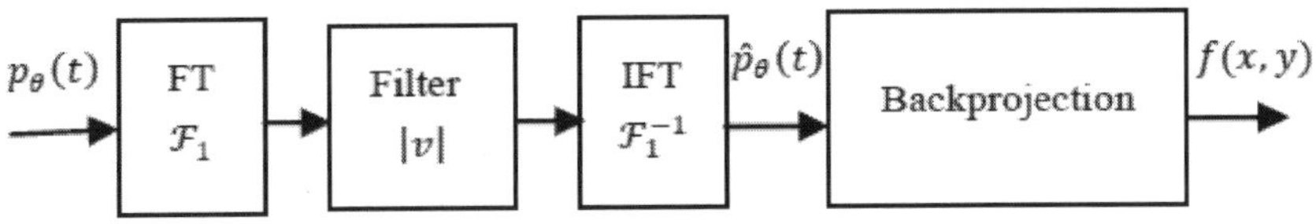

Figure 3. Filtered Back Projection algorithm

Notice that analytical reconstruction methods work well when we have a great number of projections uniformly distributed around the object and cannot give satisfactory results when the number of projections is less than thirty two as it will be shown in our simulation results:

4. ITERATIVE RECONSTRUCTION METHODS

Computed tomography using iterative reconstruction techniques is viewed as a linear inverse problem as follows: find a vector f , that is a solution of $p = Af$. So, the principle is to find a solution by successive estimates. In order to implement these algorithms, we first make an initial guess $f^{(0)}$ at the solution. Then we take successive projections. The projections corresponding to the current estimate are compared with the measured projections. The result of the comparison is used to modify the current estimate, thereby creating a new estimate. The algorithms differ in whether the correction is carried out under the form of an addition or a multiplication.

4.1. Algebraic Reconstruction Methods

The mathematical formulation of the reconstruction problem is posed as a system of linear equations [8], which should be solved when reconstructing an image:

$$\begin{array}{llll} A_{11}f_1 + & A_{12}f_2 + & A_{13}f_3 + \cdots & A_{1N}f_N = p_1 \\ A_{21}f_1 + & A_{22}f_2 + & A_{23}f_3 + \cdots & A_{2N}f_N = p_2 \\ \vdots & \vdots & \vdots & \vdots \\ A_{M1}f_1 + & A_{M2}f_2 + & A_{M3}f_3 & A_{MN}f_N = p_M \end{array} \tag{7}$$

So we write the set of linear equations for each ray as:

$$\sum_{j=1}^{N} A_{ij} f_j = p_i \,, \quad i = 1, 2, \cdots, M \tag{8}$$

Where p_i are projection data along the ith ray. In other words, the projection at a given angle is the sum of non-overlapping, equally wide rays covering the figure, M is the total number of rays (in all the projection). A_{ij} are the weights that represent the contribution of every pixel for all the different rays in the projection. f_j are the values of all the pixels in the image and N is the total number of the pixels in the image. In applications requiring a large number of views and where large-sized reconstructions are made, the difficulty with using (8) can be in the calculation, storage, and fast retrieval of the weight coefficients A_{ij}, also the dimensions of the matrix A become so large that direct factorization methods become infeasible, which is usually the case in two and three dimensions. This difficulty can pose problems in applications where reconstruction speed is important. In this case, one can use iterative methods for solving (8), which are described in brief as follows.

4.1.1. Algebraic Reconstruction Technique (ART)

The ART is sequential method, i.e., each equation is treated at a time, since each equation is dependent on the previous. The equation of ART is given by

$$f_j^{(k+1)} = f_j^{(k)} + \alpha \; \frac{p_i - \sum_{n=1}^{N} A_{in} f_n^{(k)}}{\sum_{n=1}^{N} A_{in}^2} \; A_{ij} \tag{9}$$

Where $f_j^{(k)}$ and $f_j^{(k+1)}$ are the current and the new estimates, respectively; $\sum_{n=1}^{N} A_{in} f_n^{(k)}$ is the sum weighted pixels along ray i; for the kth iteration; p_i is the measured projection for the ith ray , and α is the relaxation parameter. The second term on the left in equation (9), is the term of correction. The process starts by making an intialguess. Observing equation (9), we see that (a) this correction term is added to the current estimate in order to found the new estimate and (b) the comparison consists in the subtraction of the estimated projections from the measured projections. Also, we can easily see, that equation (9) is used to update the value of the jth pixel on every ray equation. This process can be summarized in Figure 4. Notice that the role of the relaxation parameter α is to reduce the effects of the noise which is caused by poor approximations of the ray-sums.

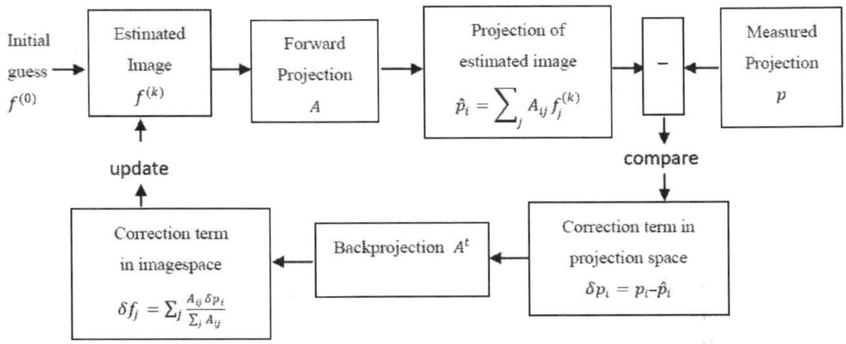

Figure 4. Algebraic Reconstruction Technique

4.1.2. Simultaneous Iterative Reconstruction Techniques (SIRT)

As the name refers to, all the methods of this class are simultaneous, which means that information from all the equations are used at the same time. In this approach, we go through all the equations, and then only at the end of each iteration are the pixel values changed, the change for each pixel being the average value of all the computed changes for that pixel. This constitutes one iteration of the algorithm. In the second iteration, we go back to the first equation in (7) and the process is repeated.

The SIRT algorithm has however not gained a wide popularity, because it has a big downside. SIRT appears to require a long time for convergence and thus takes a long time for reconstructing an image. For this reason SART or Simultaneous ART was developed.

4.1.3. Simultaneous Algebraic Reconstruction Techniques (SART)

As well known, simultaneous algebraic reconstruction technique (SART) was proposed as a major refinement and an enhanced version of ART since it is a combination of the positive aspect of SIRT and ART [8]. The equation for iterative process using SART can be represented as below:

$$f_j^{(k+1)} = f_j^{(k)} + \alpha \frac{\sum_i \left[\frac{p_i - \sum_{n=1}^{N} A_{in} f_n^{(k)}}{\sum_{n=1}^{N} A_{in}} \right] A_{ij}}{\sum_i A_{ij}} \quad (10)$$

We can easily see, from equation (10), that the correction term, is additive. The main difference between ART and SART is that the correction terms in SART are simultaneously applied for all the rays in one projection, whereas the pixel values are updated ray-by-ray in ART. This explains why SART converges faster than ART.

In the following we will discuss Gradient algorithm and Bayesian maximum a posteriori (MAP) approach. Both are optimization methods, that is, they find the best estimate for the solution fitting a given criterion; in the Gradient algorithm, the criterion is the minimization of the difference between p and Af, whereas in the Bayesian (MAP) approach, which is based on Bayes' formula as the name refers to, the criterion is the maximization of an associated objective function of the reconstructed image.

4.2. Gradient Algorithm

In iterative methods, one wants to solve $p = Af$. Where p is the vector of values in the sinogram, A is a given matrix, and f is the unknown vector of pixel values in the image to be reconstructed. One of the iterative reconstruction techniques is the Gradient algorithm. The Gradient algorithm or steepest-descent algorithm iteratively searches for f using the equation:

$$f^{(k+1)} = f^{(k)} + \alpha^{(k)} A^t \left(p - A f^{(k)}\right) \quad (11)$$

This means that the new estimate $f^{(k+1)}$ is equal to previous estimate $f^{(k)}$ plus a second term indicating the new direction (chosen to be opposite to the local gradient, and therefore directed toward the steepest-descent), weighted by a coefficient $\alpha^{(k)}$ representing the step length. This coefficient can be chosen to optimize the convergence of the process. As we can see from equation (11), the correction factor, which is the second term, is additive. The Gradient algorithm consists essentially of four steps:

- Compute $\hat{p} = Af$ (Forward projection)
- Compute $\delta p = p - \hat{p}$ (Error or residual)
- Distribute $\delta f = A^t \delta p$ (Backprojection of error)
- Update $f^{(k+1)} = f^{(k)} + \delta f$

The criterion of this optimization algorithm is the progressive minimization of the difference between the measured projections and estimated ones. That is, the error is defined as: $\|p - Af\|^2$.

In general, reconstruction problem is an ill posed inverse problem. There is immense need for prior information. However, The Gradient method, which is a deterministic regularization process, does not have enough tools to account for stronger prior information needed in some applications. For this reason, probabilistic methods have been developed. One of these probabilistic methods is Bayesian approach which will be described in the following.

4.3. Bayesian Approach

The goal is to find f with respect to two requirements: (a) estimated projections have to be as close as possible to measured projections and (b) the reconstructed images have to be not too noisy. Therefore, the introduction of a prior knowledge as constraint, may considerably favor convergence of iterative algorithms. This process is called regularization. The prior, based on an assumption of what the true image is, is usually chosen to penalize the noisy images. We consider the linear model with additive white Gaussian noise, i.e.,

$$p = Af + n \tag{12}$$

Where p, f and n are vectors of random variables. In Bayesian inversion theory, the complete solution for an inverse problem is represented by the posterior distribution, given by Bayes' formula

$$p(f/p) = \frac{p(f)p(p/f)}{p(p)} \propto p(f)p(p/f) \tag{13}$$

Where $p(p/f)$ is the likelihood density, $p(f)$ is the prior density and $p(p)$ is normalization constant. We select the maximum a posteriori (MAP) estimate which is obtained from

$$\hat{f} = \mathrm{argmax}_f \, p(f/p) \tag{14}$$

It means that for the given prior density $p(f)$ and the measurement data p, we determine the unknown values f which are in the best agreement with the model (12). Assuming zero mean, isotropique Gaussian noise with variance σ_n^2, the likelihood function is

$$p(p/f) \propto \exp\left(-\frac{\|p - Af\|^2}{2\sigma_n^2}\right) \tag{15}$$

The next question is how to choose the prior density function $p(f)$. For the sake of clarity, we select the Gaussian white noise prior with the positivity constraint, i.e.,

$$p(f) \propto \exp\left(-\frac{\|f\|^2}{2\sigma_f^2}\right) u(f) \tag{16}$$

Where the step function $u(f)$ equals to one when all the elements in f are positive; otherwise it is zero. The computation of MAP implies minimizing

$$\hat{f} = \arg\min_f \left\{ \|p - Af\|^2 + \lambda \|f\|^2 \right\} \tag{17}$$

Where the regularization parameter $\lambda = \sigma_n^2 / \sigma_f^2$. The higher the noise level, the larger the regularization parameter λ. The statistical estimation problem is hence converted into an optimization problem. In other words, at each iteration k, a current estimate of the image is available. Using a system model (which may include attenuation and blur), it is possible to simulate what projections of this current estimate should look like. The measured projections are then compared with simulated projections of the current estimate, and the error between these simulated and measured projections is used to modify the current estimate to produce an updated (and hopefully more accurate) estimate, which becomes iteration $k+1$. This process is then repeated many times. In practice, the MAP-estimate is computed by minimum norm least square (MNLS) iterative method implemented in Matlab Optimization Toolbox. But, Bayesian approach is not only limited to MAP estimator. Many other estimators have been proposed such as the Posterior mean and Marginal MAP estimator. So, the main steps in Bayesian approach are (a) prior density modeling (which can be Gaussian, Generalized Gaussian, Gamma mixture, etc. ... .) and (b) the choice of the estimator and computational aspects. The most important advantage of Bayesian approach is that it is a coherent approach to combine information content to the data and priors.

5. RECONSTRUCTION QUALITY, IN THE CONTEXT OF COMPUTATIONAL PERFORMANCE

The BP, FBP, Gradient and Bayesian algorithms are implemented and tested for two test images. These algorithms have been implemented on a PC using Matlab programming language. In order to objectively evaluate the reconstructed results, we have computed four image quality measurement parameters proposed in [3] in addition to two relative norm errors and the visual quality of the reconstructed image. The quality measurements are listed below:

The relative norm error of the resulting images [8] is used and defined as:

$$df = \frac{\|f-\hat{f}\|^2}{\|f\|^2} \quad (18)$$

Where f is the gray level value of the test image and $\hat{f}$ is the gray level value of the reconstructed image. The relative norm error of the simulated projections and defined as

$$dp = \frac{\|p-\hat{p}\|^2}{\|p\|^2} \quad (19)$$

Where p is the measured projection and $\hat{p}$ is the simulated projection.

Mean square error (MSE): it's measure between the test image and the reconstructed image and defined as:

$$MSE = \frac{1}{N}\sum_{j=1}^{N}\left[f_j - \hat{f}_j\right]^2 \quad (20)$$

Where f_j is the value of the pixel in the test image $\hat{f}_j$ and the value of the pixel in the reconstructed image, N is the total number of pixels.

Normalized cross-correlation (NCC): is defined as:

$$NCC = \sum_{j=1}^{N}\left[f_j - \hat{f}_j\right]^2 / \sum_{j=1}^{N}\left[f_j\right]^2 \tag{21}$$

Smaller errors df, dp, MSE and NCC, means that the resulting reconstructed image is closer to the test image.

Structural content (SC): it's the measure of image similarity based on small regions of the images containing significant low level structural information and is defined as:

$$SC = \sum_{j=1}^{N} f_j^2 / \sum_{j=1}^{N} \hat{f}_j^2 \tag{22}$$

Peak Signal to Noise Ratio $(PSNR)$: is defined as:

$$PSNR = 10\log\left(\frac{255}{\sqrt{MSE}}\right) \tag{23}$$

Another criterion for the iterative reconstruction techniques is the number of iterations. Smaller iteration is preferable. Projections (parallel beam type) for the image reconstruction are calculated analytically by defining the first test image: Shepp-Logan phantom head model (Simulated) image as it is shown in Figure 2. Figure 5 shows the second test image which is the standard medical image of abdomen. Projections are calculated mathematically. The two original test images are grayscale images of size 128×128 and 256×256 respectively, with coverage angle ranging from 0° to 180°with rotational increment of 10° to 2°.

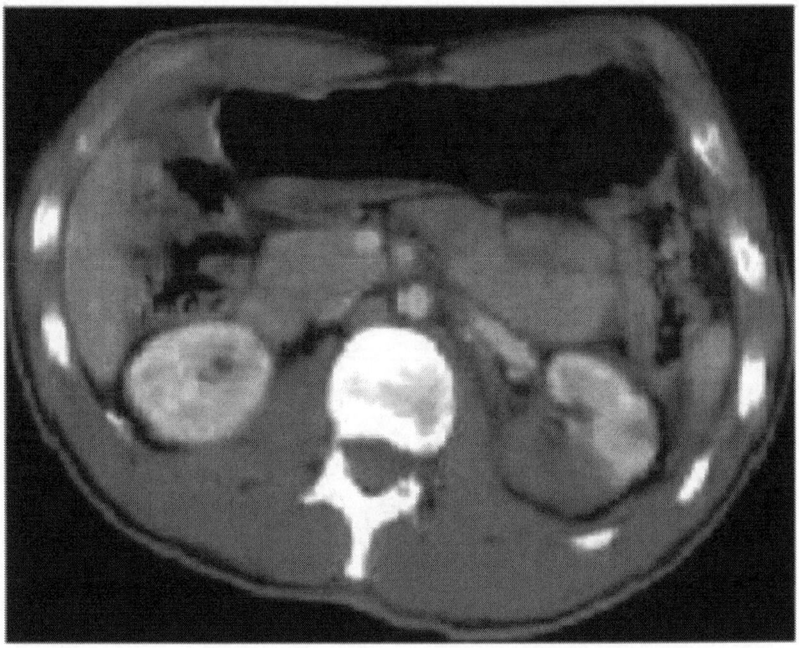

Figure 5. Input test image: Standard medical image of abdomen

6. RESULTS AND DISCUSSION

Comparison of reconstruction techniques such as BP, FBP (analytical), Gradient and Bayesian algorithm (iterative) with respect to performance parameters and visual quality of reconstructed images is presented in this section. The reconstruction performances are calculated versus number of projections and number of iterations. Figure 6 shows the reconstruction of 2D Shepp-Logan phantom head model (Figure 2) by BP with different projections. The simulated Peak Signal to Noise Ratio PSNR results as well as Figure 7 show that minimum 64 projections, with coverage angle ranging from 0°to 180° with an incremental value of 3° is necessary to reconstruct the image with acceptable quality. This technique generates star or spoke artifact. Figures 6 and 7 clearly reveal that the quality of reconstructed image slightly increases as number of projections increases. The reconstructed images appear to be very blurry. The quality measurements versus projections of Shepp-Logan image using BP are drawn in Figure 8. The later reveals very close similarity among computational performance with the visual quality of reconstructed image shown in Figure 6.

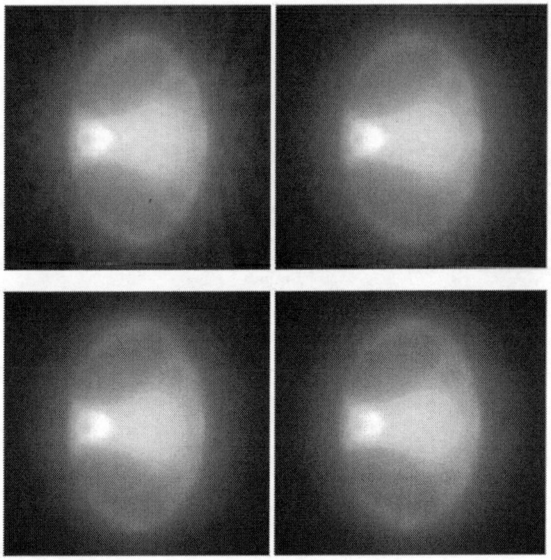

Figure 6. Reconstructed image of Shepp-Logan phantom by BP. Using (a) 16, (b) 32, (c) 64 and (d) 180 number of projections (from left to right, up to bottom)

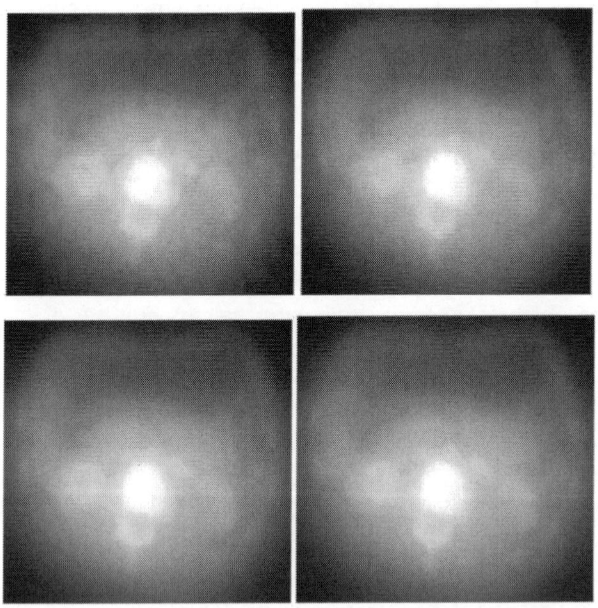

Figure 7. Reconstructed image of standard medical image by BP using 16, 32, 64 and 180 number of projections (from left to right, up to bottom)

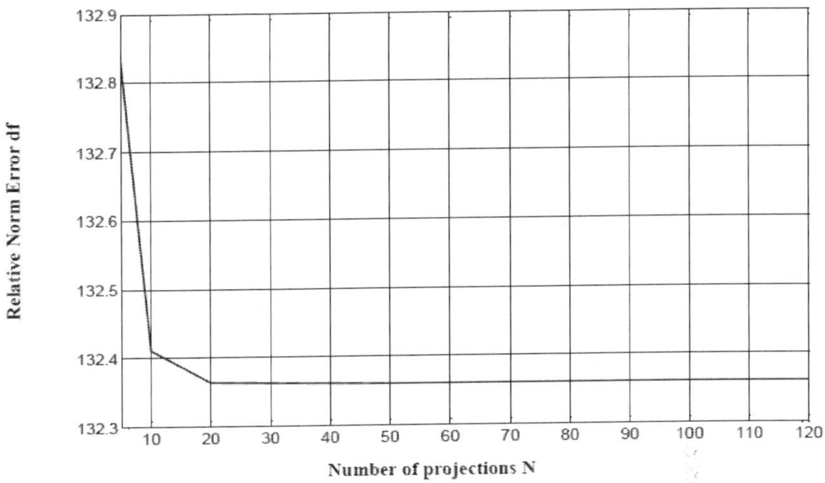

132.9
132.8
132.7
132.6
132.5
132.4
132.3
Relative Norm Error df
10
20
30
40
50
60
70
80
90
100
110
120
Number of projections N

(a)

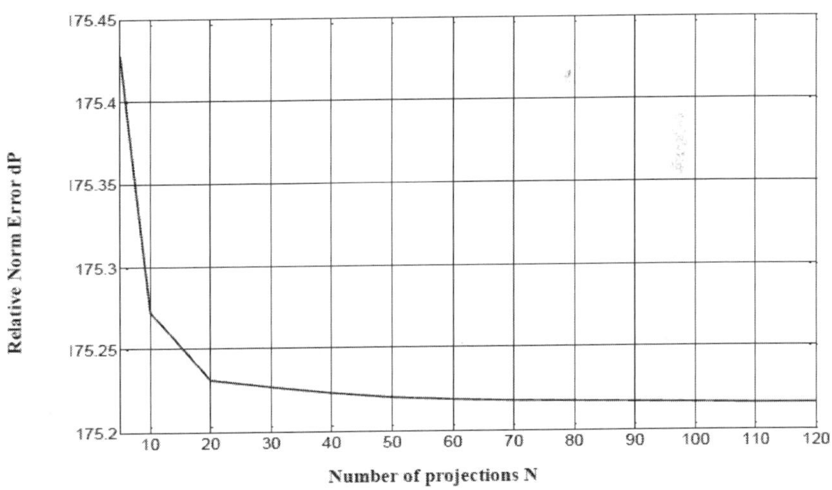

175.45
175.4
175.35
175.3
175.25
175.2
Relative Norm Error dP
10
20
30
40
50
60
70
80
90
100
110
120
Number of projections N

(b)

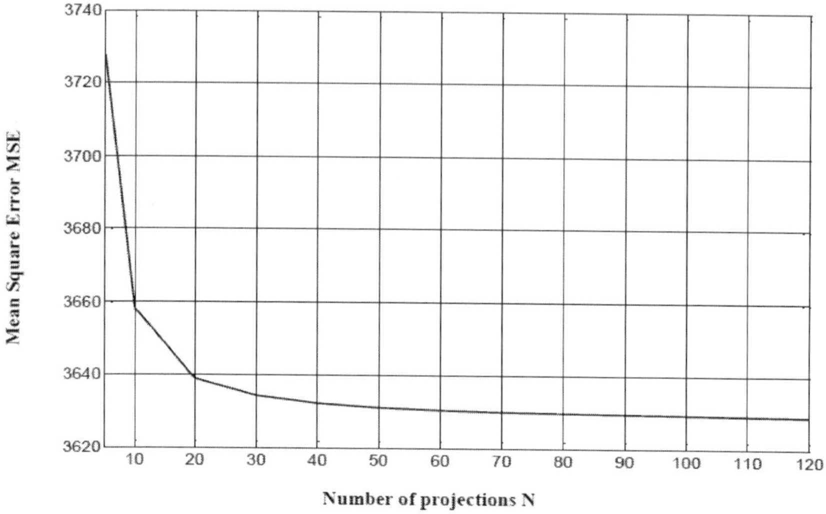
3740
3720
3700
3680
3660
3640
3620
Mean Square Error MSE
10
20
30
40
50
60
70
80
90
100
110
120
Number of projections N

(c)

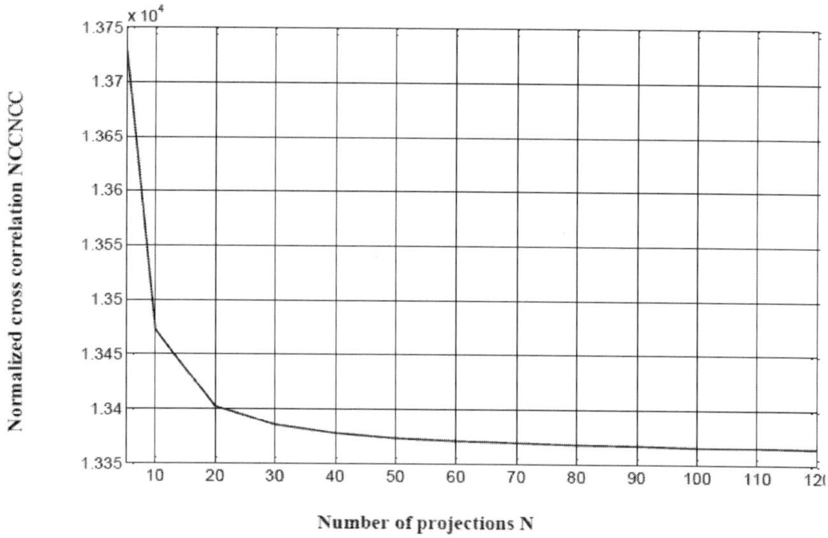
x 10^4
1.375
1.37
1.365
1.36
1.355
1.35
1.345
1.34
1.335
Normalized cross correlation NCCNCC
10
20
30
40
50
60
70
80
90
100
110
Number of projections N

(d)

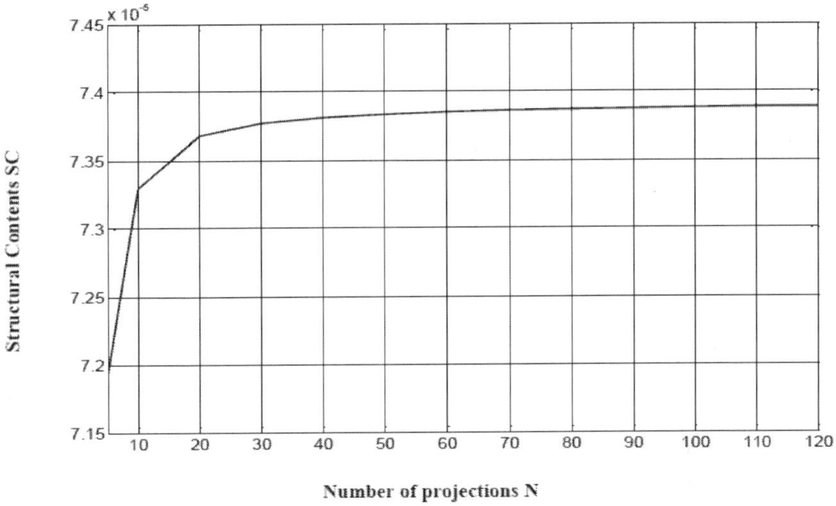

(e)

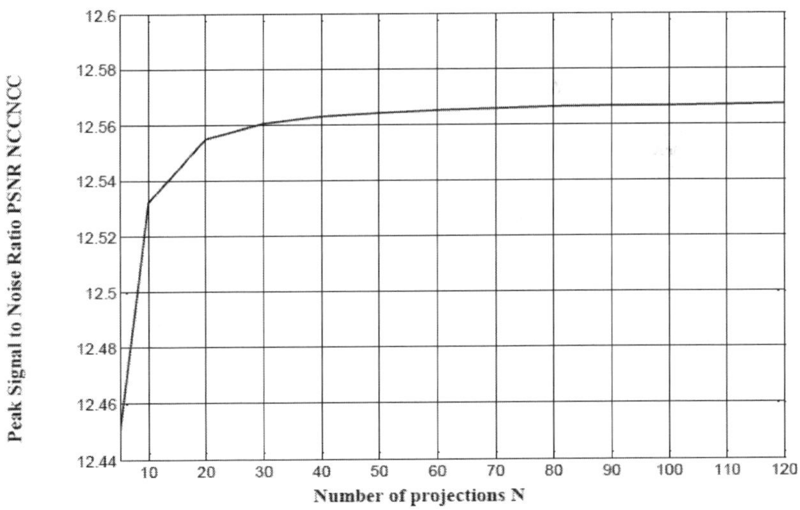

(f)

Figure 8. Quality measurements v/s number of projections for BP. (**a**) *df* (**b**) *dp* (**c**) *MSE* (**d**) *NCC* (**e**) *SC* (**f**) *PSNR* .

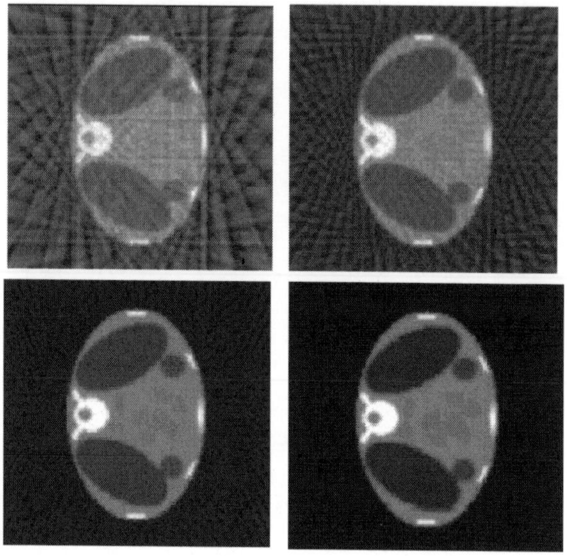

(a)

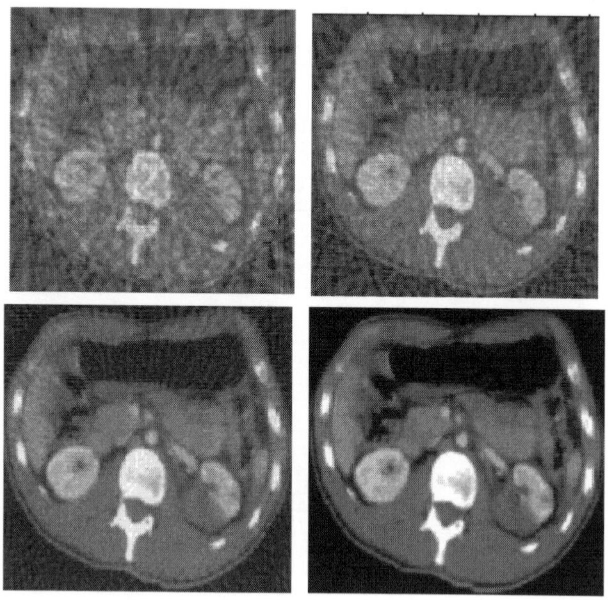

(b)

Figure 9. Reconstructed image of (a) Shepp-Logan phantom (b) standard medical image, by FBP (with Shepp-Logan filter).Using 16, 32, 64 and 180 number of projections (from left to right)

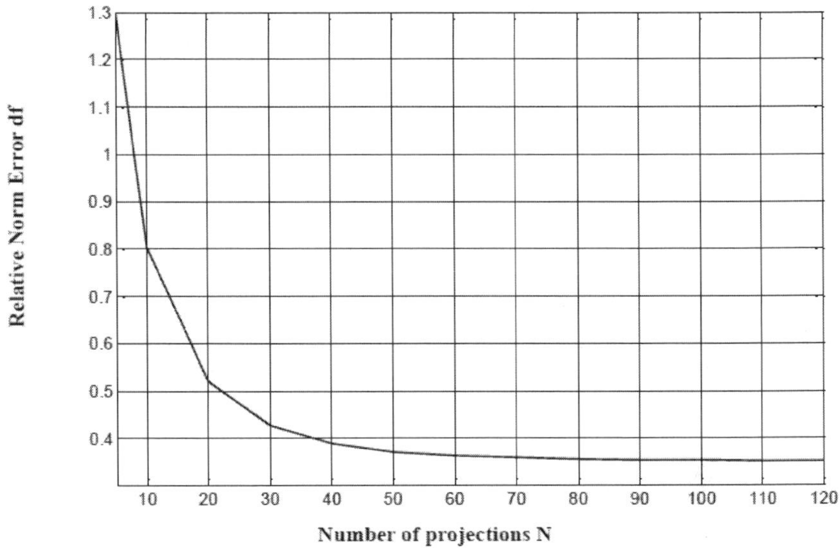
Relative Norm Error df
Number of projections N
1.3
1.2
1.1
1
0.9
0.8
0.7
0.6
0.5
0.4
10
20
30
40
50
60
70
80
90
100
110
120

(a)

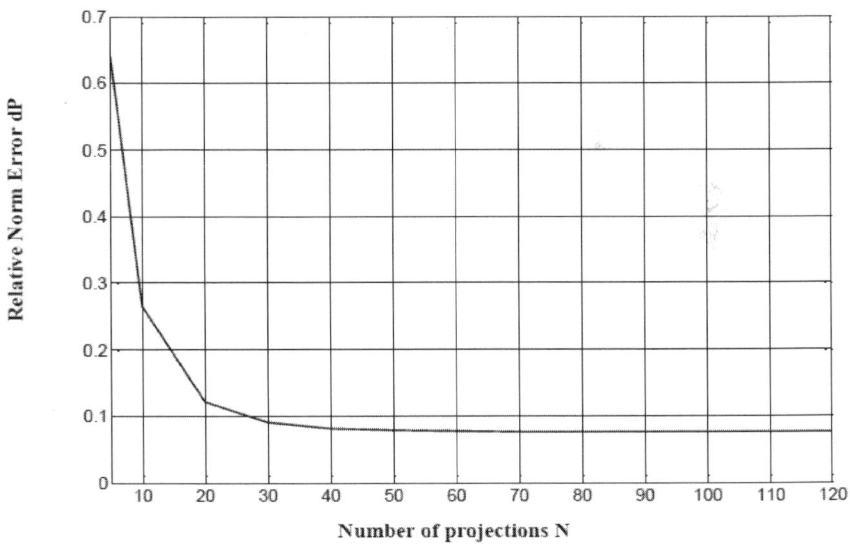
Relative Norm Error dP
Number of projections N
0.7
0.6
0.5
0.4
0.3
0.2
0.1
0
10
20
30
40
50
60
70
80
90
100
110
120

(b)

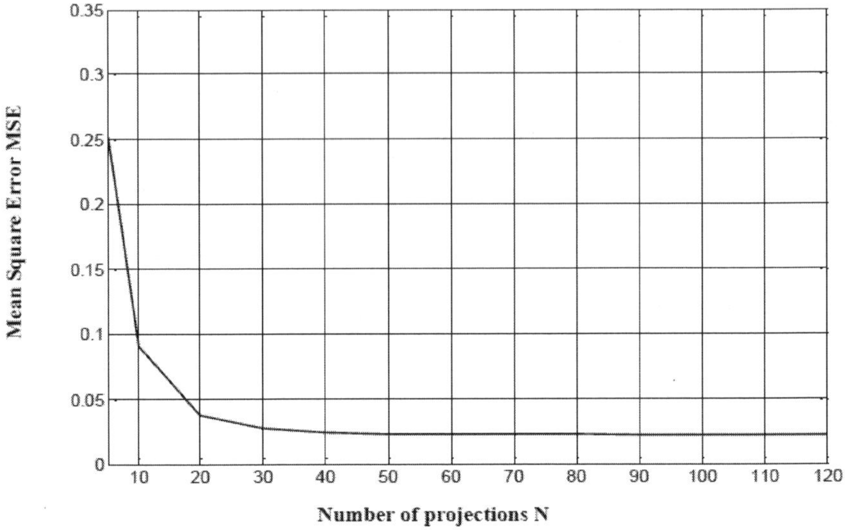
0.35
0.3
0.25
0.2
0.15
0.1
0.05
0
10
20
30
40
50
60
70
80
90
100
110
120
Mean Square Error MSE
Number of projections N

(c)

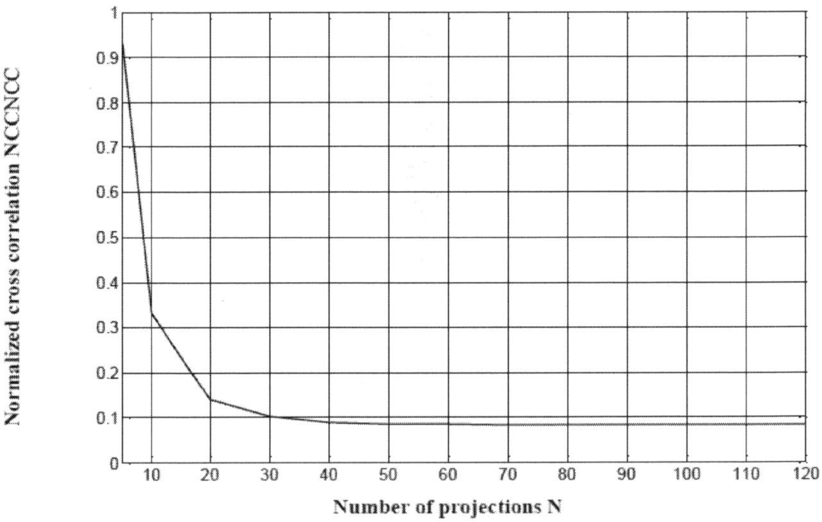
1
0.9
0.8
0.7
0.6
0.5
0.4
0.3
0.2
0.1
0
10
20
30
40
50
60
70
80
90
100
110
120
Normalized cross correlation NCCNCC
Number of projections N

(d)

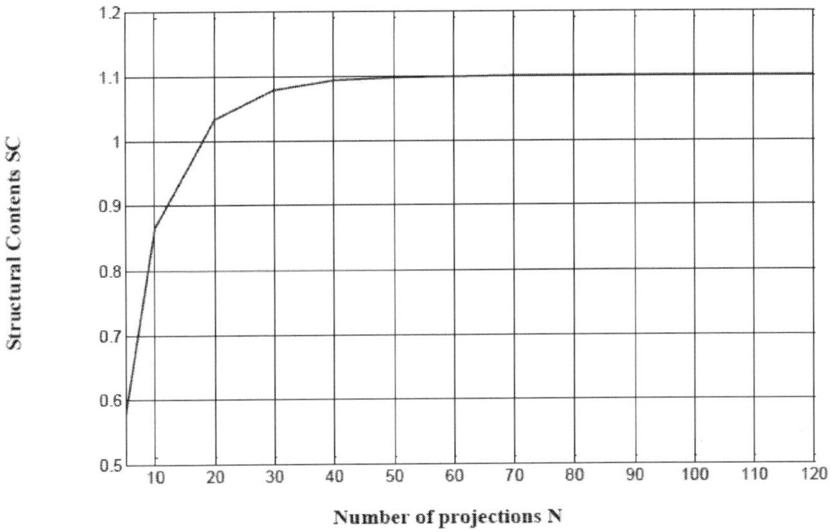

(e)

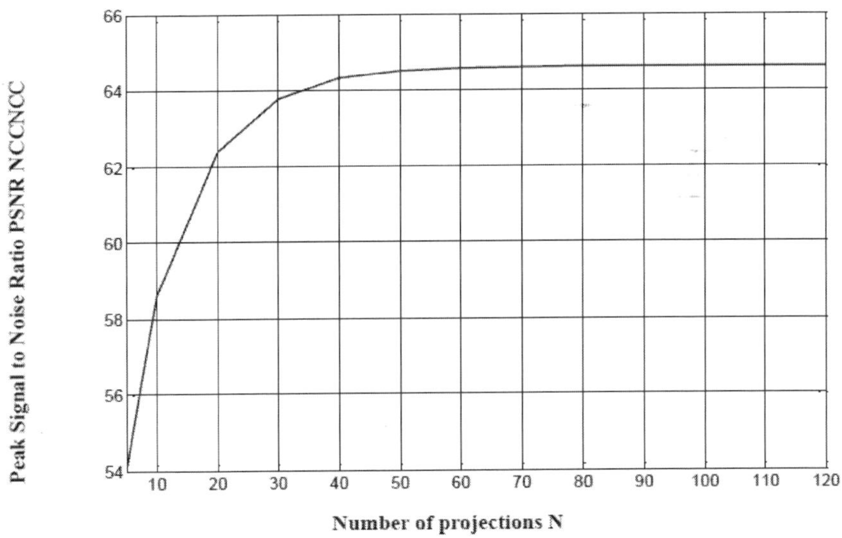

(f)

Figure 10. Quality measurements v/s number of projections for FBP with Hanning filter. (a) *df* (b) *dp* (c)*MSE* (d) *NCC* (e) *SC* (f) *PSNR* . Test image: Shepp-Logan.

Test image:

Shepp-Logan

Figure 9 shows the reconstruction of phantom head model and the standard medical of abdomen by FBP with coverage angle ranging from 0°to 180°with an incremental value 10° of to 2°. Figure 9 clearly reveals that the quality of reconstructed images of both Shepp-Logan phantomand the standard medical image of abdomen increases as number of projections increases. The graphical representation of quality measurements versus projections of Shepp-Logan test Image for FBP is shown in Figure 10.

We can easily observe that the performance parameters remain almost constant after 32 projections. This algorithm requires a minimum of about 32 projections with rotational increment of 5° to display acceptable reconstructed image. It is the most common method of removing the star artifact which is normally generated in BP by using various filters. FBP algorithm is fast and efficient with large number of projections. Also after about 32 projections, trend of errors appear to decrease consistently, unlike $PSNR$ which significantly increases after 32 projections. Table 1 shows the quality measurements by varying the filter used in FBP and keeping common number of projections.

Table 1. Quality measurements by varying the filter keeping fixed number of projections (32)

FBP					
Test image: Sheep Logan					
Filter	Rampe	Hanning	Hamming	Cosine	Shepp - Logan
df	0.4168	0.3584	0.3616	0.3761	0.4008
dp	0.0880	0.0789	0.0795	0.0821	0.0858
MSE	0.0359	0.0266	0.0270	0.0292	0.0332
PSNR	62.6048	63.9145	63.8382	63.4966	62.9447
NCC	0.1324	0.0979	0.0997	0.1078	0.1225
SC	1.0235	1.0814	1.0777	1.0595	1.0371
MD	1.3833	1.2730	1.2755	1.3105	1.3494
NAE	0.4427	0.3367	0.3430	0.3710	0.4151

Table 1 demonstrates that the quality measurements, from FBP using Hanning filter are slightly better than those of other filters. This means that the choice of the filter has not a significant effect to enhance the image quality. From Figure 11, it can be clearly seen that visual quality of the reconstructed images for various filters is almost similar.

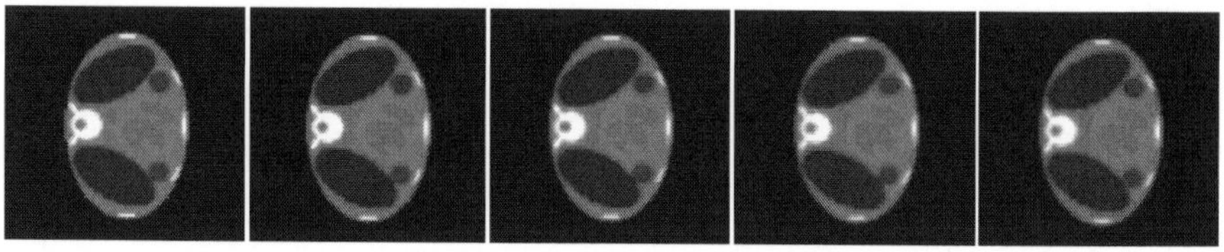

Figure 11. Reconstructed images by FBP. Using (a) Ramp, (b) Shepp-Logan, (c) Cosine, (d) Hamming and (e) Hanning filters, number of projections 180

The resultant reconstructed images obtained from Gradient algorithm by varying the number of iterations, are shown in Figure 12. The later demonstrates that Gradient algorithm is providing better reconstruction than that of FBP and BP. The quality measurements versus projections of Shepp-Logan image using Gradient algorithm for 200 iterations are shown in Figure13. We can easily observe that performance parameters are considerably enhanced comparatively with those of FBP especially for low number of projections, since error values obtained from Gradient algorithm are significantly reduced with a significant increase of $PSNR$. We should note here that the number of iterations is much required in order to enhance the image quality.

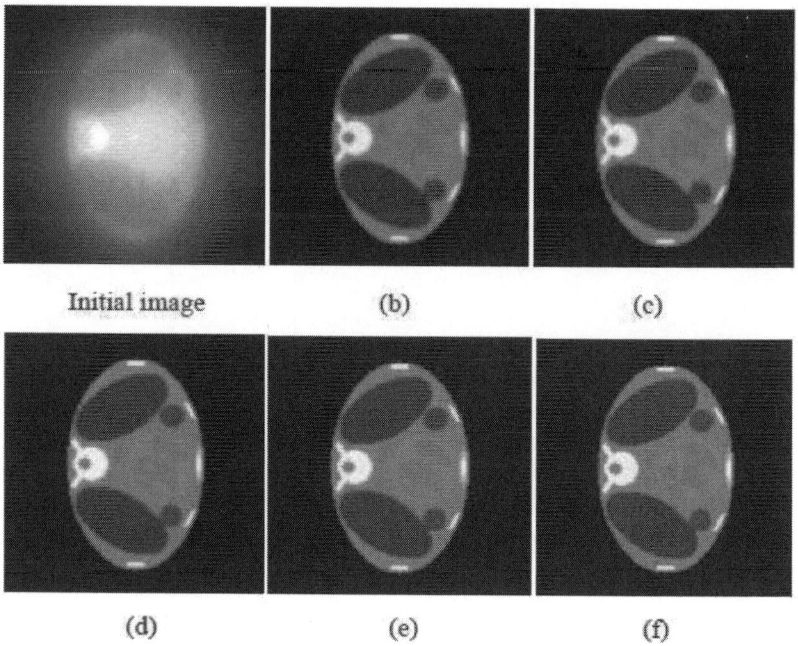

Figure 12. Reconstructed images by Gradient algorithm. Using (b) 20, (c) 60, (d) 100 (e) 200 and (f) 500 iterations

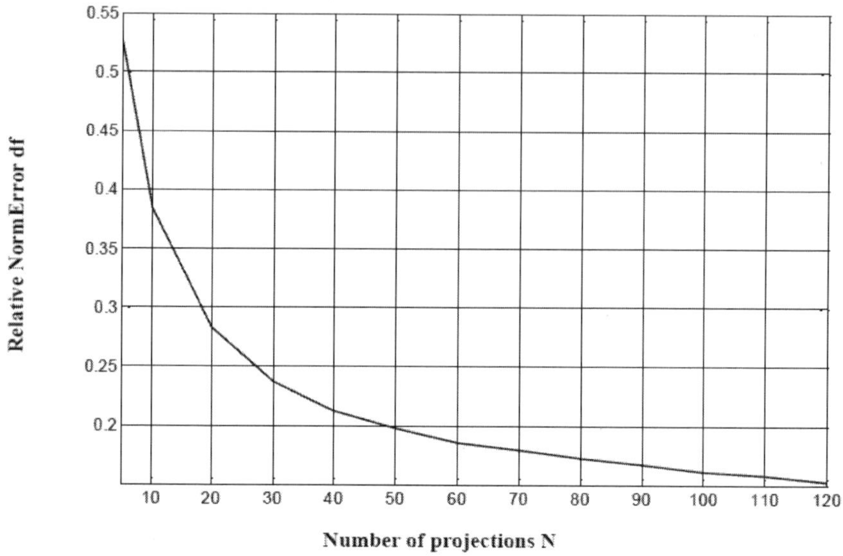
0.55
0.5
0.45
0.4
0.35
0.3
0.25
0.2
Relative NormError df
10 20 30 40 50 60 70 80 90 100 110 120
Number of projections N

(a)

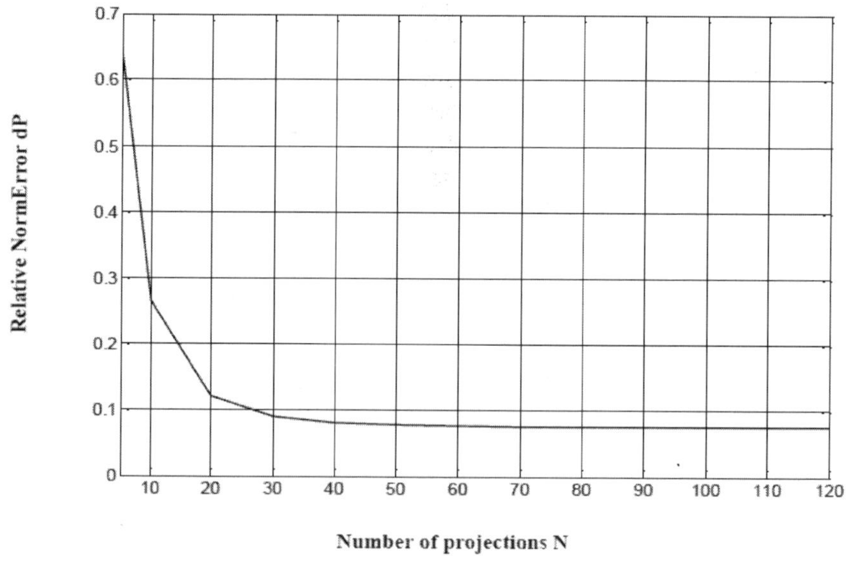
0.7
0.6
0.5
0.4
0.3
0.2
0.1
0
Relative NormError dP
10 20 30 40 50 60 70 80 90 100 110 120
Number of projections N

(b)

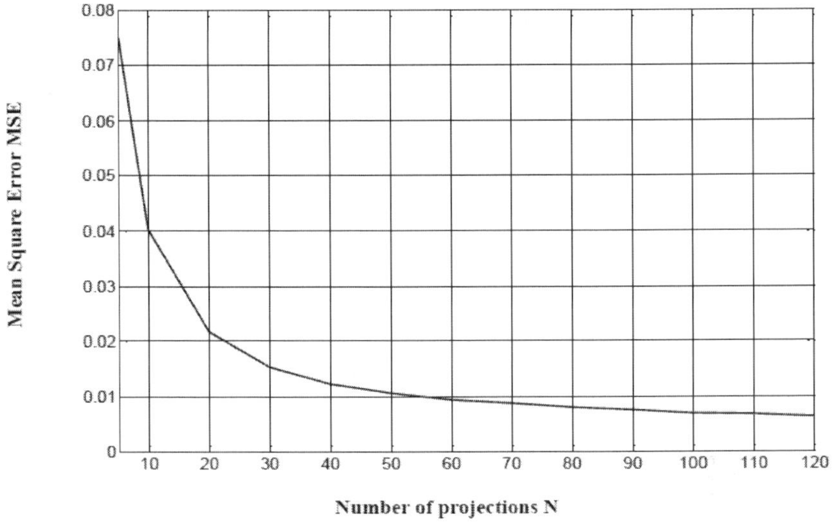
Mean Square Error MSE
0.08
0.07
0.06
0.05
0.04
0.03
0.02
0.01
0
10 20 30 40 50 60 70 80 90 100 110 120
Number of projections N

(c)

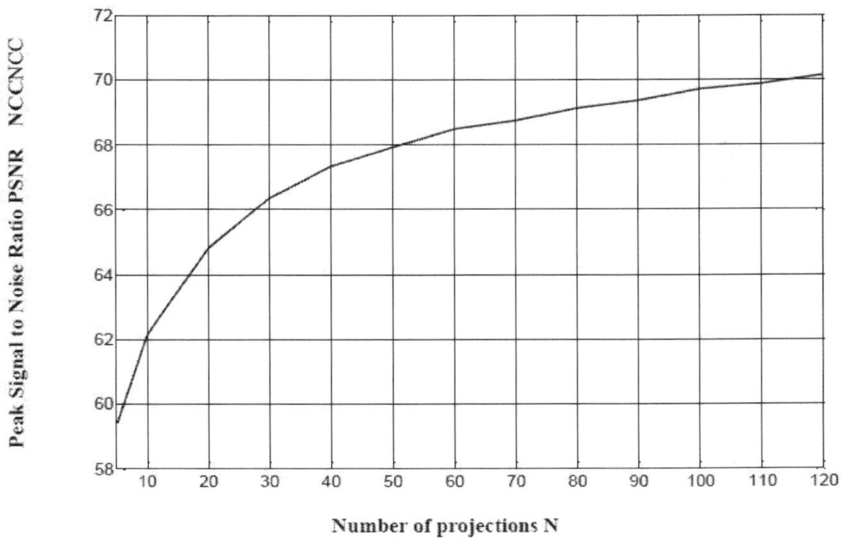
Peak Signal to Noise Ratio PSNR NCCNCC
72
70
68
66
64
62
60
58
10 20 30 40 50 60 70 80 90 100 110 120
Number of projections N

(d)

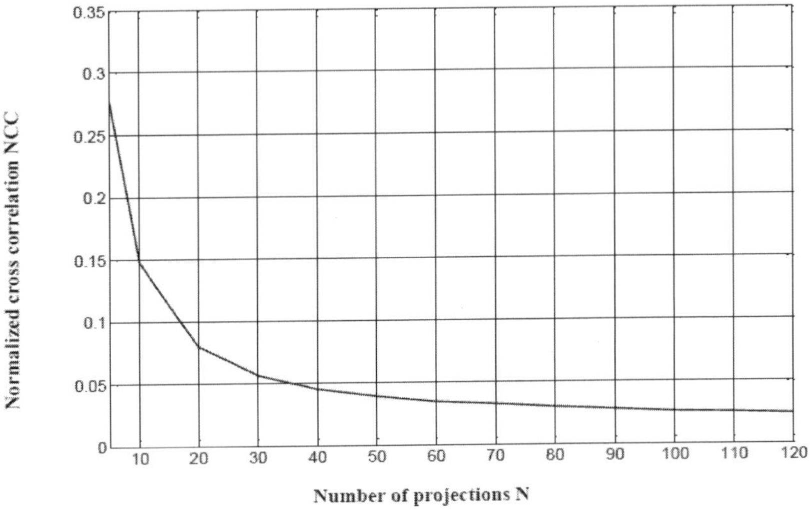

(e)

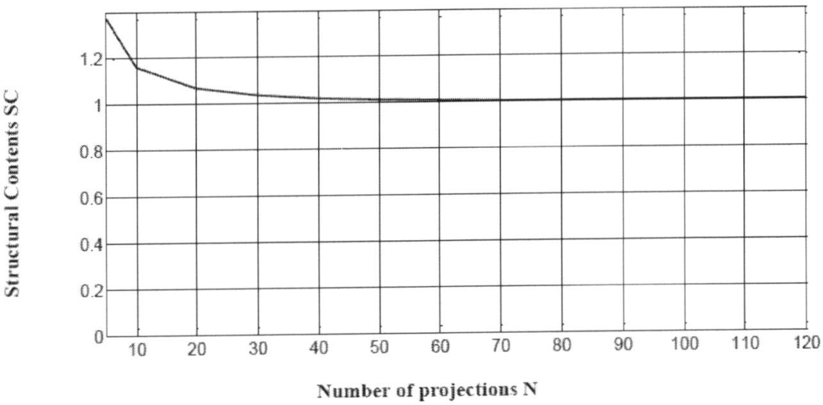

(f)

Figure 13. Quality measurements v/s number of projections for Gradient algorithm. Using 200 iterations (a) df (b) dp (c) MSE (d) $PSNR$ (e) NCC (f) SC. Test image: Shepp-Logan

Figure 14 shows the resultant reconstructed images obtained from Bayesian algorithm by varying the number of iterations. Figure 15 shows the quality measurements versus projections, obtained from Bayesian algorithm keeping common number of iterations $(k = 200)$. The experiments reveal the fact that Bayesian algorithm effectively eliminated Star artifacts created by BP. The quality measurements are significantly enhanced comparatively with those of Gradient and FBP techniques.

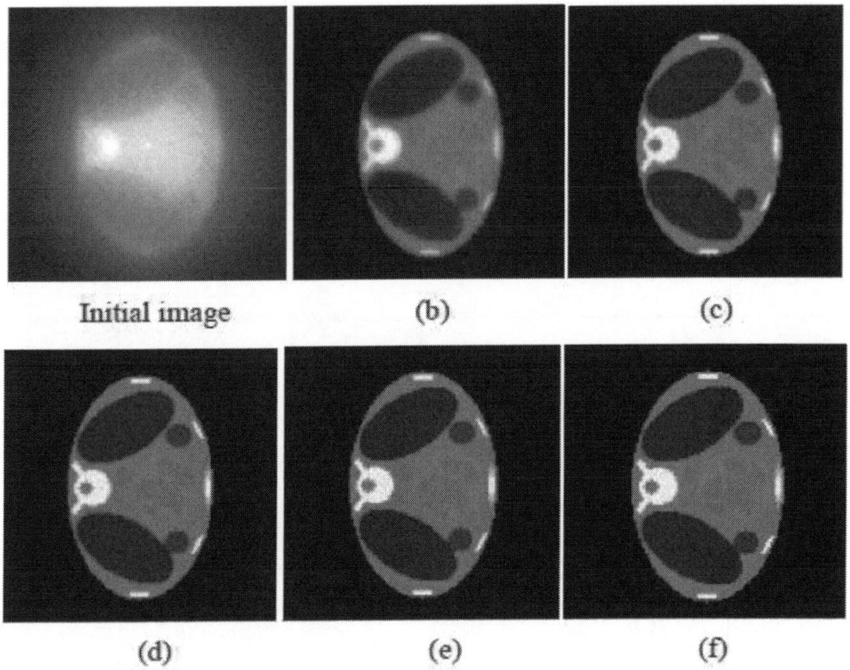

Figure 14. Reconstructed images by Bayesian algorithm using (b) 20, (c) 60, (d) 100, (e) 200 and (f) 500 iterations

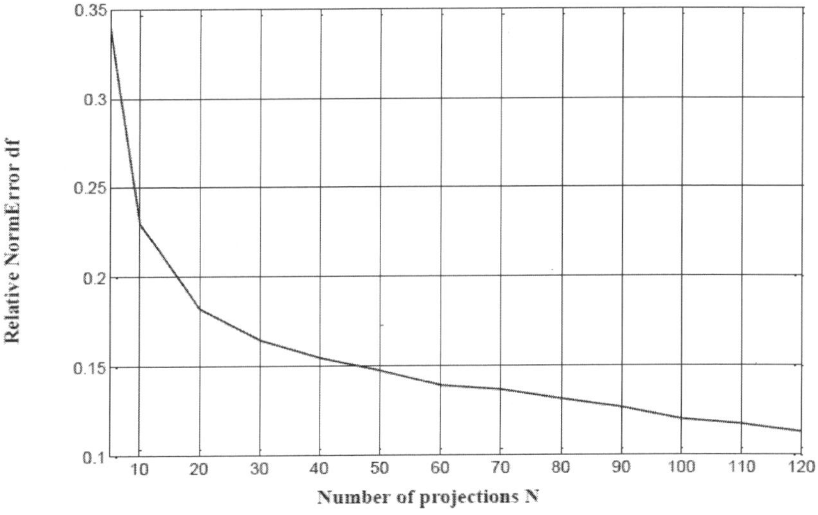

(a)

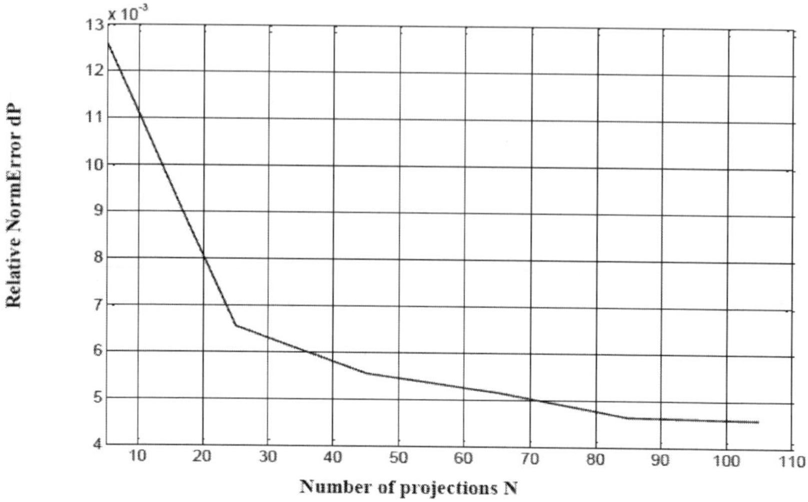
x 10^-3
Relative NormError dP
Number of projections N

(b)

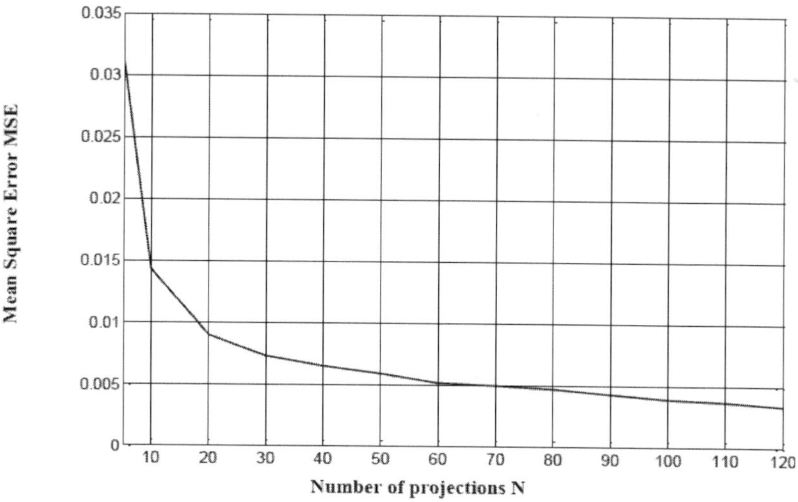
Mean Square Error MSE
Number of projections N

(c)

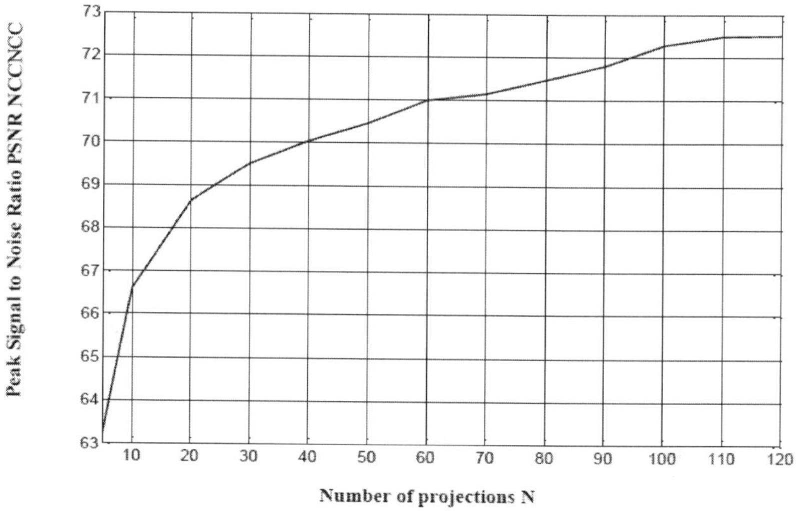
Peak Signal to Noise Ratio PSNR NCCNCC
73
72
71
70
69
68
67
66
65
64
63
10 20 30 40 50 60 70 80 90 100 110 120
Number of projections N

(d)

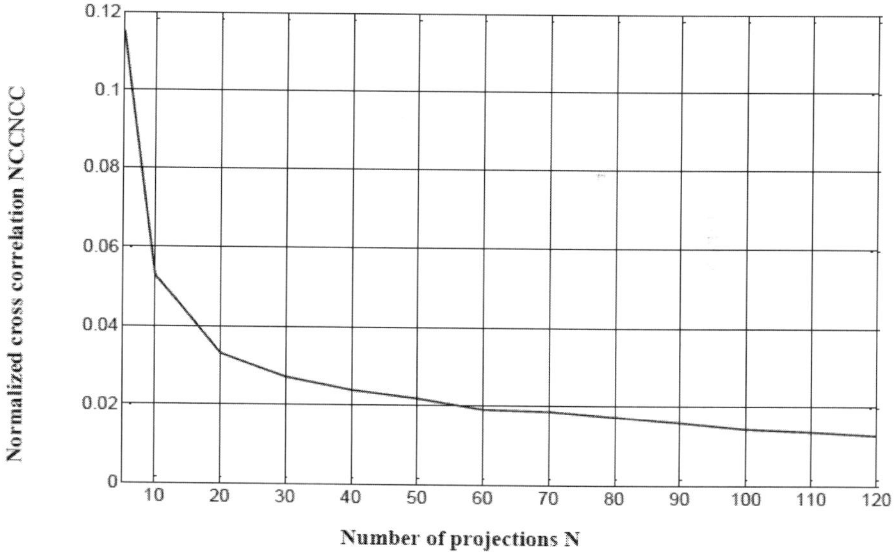
Normalized cross correlation NCCNCC
0.12
0.1
0.08
0.06
0.04
0.02
0
10 20 30 40 50 60 70 80 90 100 110 120
Number of projections N

(e)

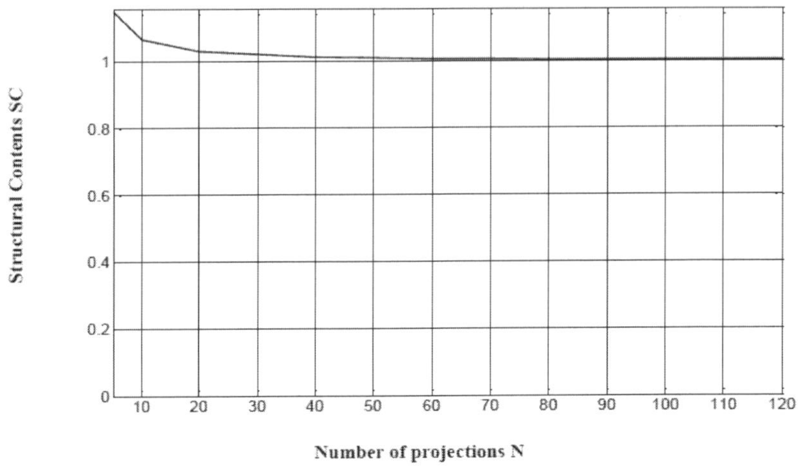

(f)

Figure 15. Quality measurements v/s number of projections for Bayesian algorithm. Using 200 iterations (a) *df* (b) *dp* (c) *MSE* (d) *PSNR* (e) *NCC* (f) *SC* . Test image: Shepp-Logan

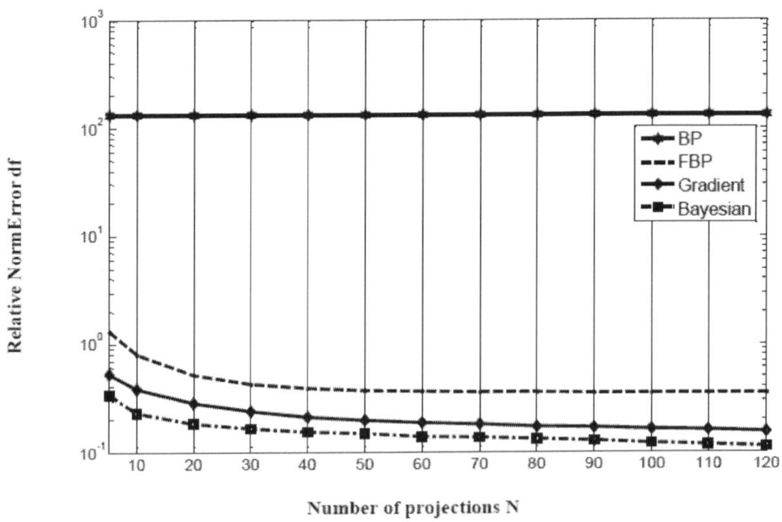

(a)

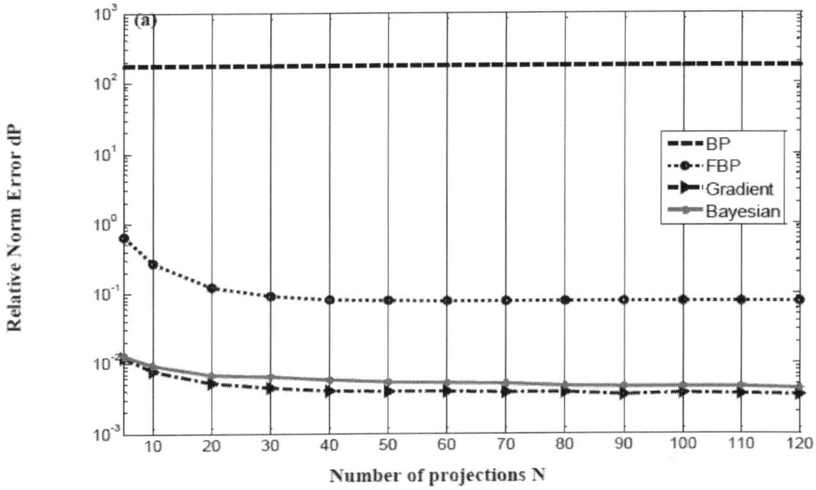
(a)
Relative Norm Error dP
BP
FBP
Gradient
Bayesian
Number of projections N

(b)

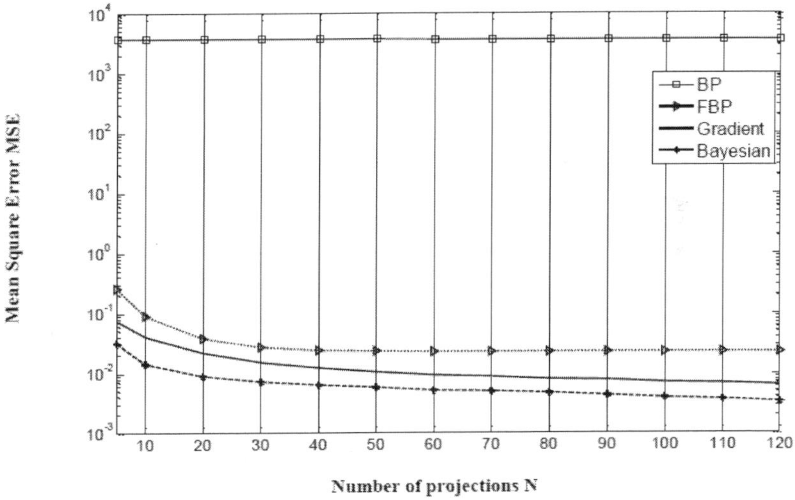
Mean Square Error MSE
BP
FBP
Gradient
Bayesian
Number of projections N

(c)

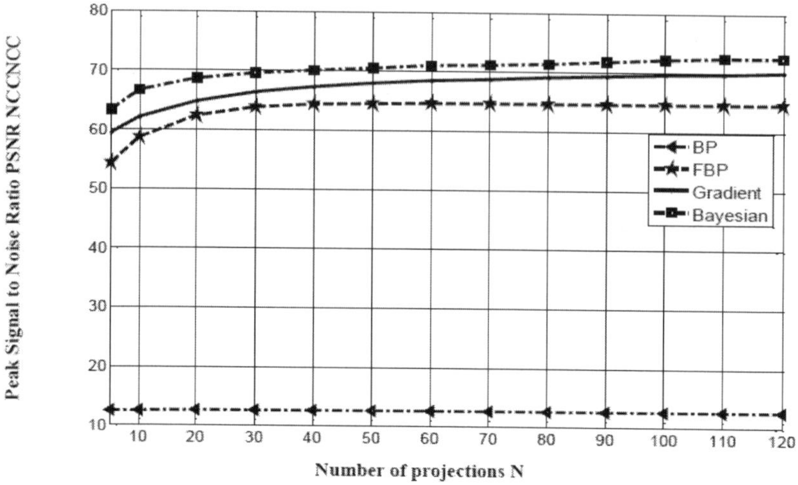

(d)

Figure 16. Comparison of reconstruction techniques. In term of (**a**) *df* (**b**) *dp* (**c**) *MSE* (**d**) *PSNR* with different projections and common number of iterations for Gradient and Bayesian algorithms (200). Test image: Shepp-Logan

To more illustrate the benefits of the Bayesian approach, we provide Figure 16, Tables 2 and 3 which compare the quality measurement parameters obtained via reconstruction techniques discussed in this work. Table 4 shows the difference between Gradient and Bayesian algorithms of Shepp-Logan phantom in term of performance parameters using 100 and 200 iterations. Bayesian algorithm performs better even at limited number of projections as shown in Figure 16, and has better quality of reconstruction.

The experiments revealed major observations; as the number of projections within a given angular range was increased, the quality of reconstructed image appeared better for both analytical and iterative algorithms. The errors df, dp, MSE and NCC obtained from the Bayesian algorithm are less than those of BP, FBP and Gradient algorithms. A high value of $PSNR$ is obtained via Bayesian algorithm. Also, the star artifact appearing in image reconstructed from the Filtered Back Projection is disappeared with the method of Gradient and Bayesian algorithms. However, Gradient and Bayesian algorithms take more time to complete the process than does FBP. But for smaller number of projections, time is almost equal for both Gradient and Bayesian algorithms. Results obtained using Mat lab (The Math Works, Inc., Natick, MA), version 7.10 (R2010a), under the Window XP operating system (Microsoft Corp., Redmond, WA), on Intel Pentium Core2 Duo, CPU 3 GHz, 1.93

GB RAM. It was found, for both analytical and iterative methods studied in this work that the quality measurements except SC and $PSNR$, decreased with increasing number of projections. We can clearly observe, from Tables 2, 3 and 4, that lower values for MSE, which it means lesser error, correspond to high values of $PSNR$. Logically, a higher value of $PSNR$ is good because it means that the ratio of Signal to Noise is higher. Here, the 'signal' is the test image, and the 'noise' is the error in reconstruction. The number of iterations in the case of Gradient and Bayesian algorithms is much required in order to enhance the image quality.

Table 2. Performance parameters by varying number of projections for reconstruction techniques. Test image: Shepp-Logan

Performance Parameters	BP				FBP			
	3 Projections	16 Projections	32 Projections	64 Projections	3 Projections	16 Projections	32 Projections	64 Projections
df	136.1298	132.6530	132.478	132.412	1.6172	0.4940	0.3584	0.3319
dp	173.4889	174.4930	174.811	175.001	1.1435	0.1249	0.0789	0.0742
MSE	38370	3643.4600	3633.88	3630.28	0.5415	0.0505	0.0266	0.0228
$PSNR$	12.3249	12.5497	12.5611	12.5654	50.8285	61.1284	63.9145	64.5815
NCC	14131	13418.7000	13383.4	13370.1	1.9944	0.1861	0.0979	0.0840
SC	6.9888e-005	7.3586e-005	7.37817e-005	7.38561e-005	0.3378	0.9860	1.08145	1.0985

Table 3. Performance parameters by varying number of projections for reconstruction techniques. Test image: Shepp-Logan

Performance Parameters	Gradient				Bayesian			
	200 Iterations							
	3 Projections	16 Projections	32 Projections	64 Projections	3 Projections	16 Projections	32 Projections	64 Projections
df	0.6056	0.3177	0.2323	0.1834	0.4831	0.1974	0.1584	0.1383
dp	0.0143	0.0054	0.0043	0.0042	0.0149	0.0076	0.0059	0.0049
MSE	0.0996	0.0274	0.0147	0.0091	0.0634	0.0106	0.0068	0.0052
$PSNR$	58.1818	63.7843	66.5048	68.5572	60.1448	67.9161	69.8295	71.0056
NCC	0.3669	0.1010	0.0540	0.0336	0.2335	0.0390	0.0251	0.0191
SC	1.5762	1.0962	1.0350	1.0084	1.3156	1.0395	1.0183	1.0075

Table 4. Performance parameters by varying number of iterations for Gradient and Bayesian approach. Test image: Shepp-Logan

Performance Parameters	Gradient		Bayesian	
	100 Iiterations	200 Iterations	100 Iterations	200 Iterations
df	0.1308	0.1204	0.0828	0.0628
dp	0.0017	0.0010	0.0212	0.0006
MSE	0.0047	0.0039	0.0019	0.0011
$PSNR$	71.4887	72.2120	75.4590	77.8676
NCC	0.1740	0.0145	0.0069	0.0039
SC	1.0296	1.0225	1.0095	1.0098

7. CONCLUSIONS

This paper presents the comparisons of the image reconstruction algorithms using BP, FBP analytical methods, Gradient and Bayesian iterative algorithms. In this work, objective measurement by six performance parameters, led to an ability to subjectively judge the reconstructed image quality. By Gradient and Bayesian algorithms, the problems in star artifacts, the quality measurements and the reconstructed image quality can be significantly improved. The results show that Bayesian method provides the best image quality, small values of the errors df, dp, MSE and NCC and high values of $PSNR$ especially for large number of iterations. From the simulated results, we shall conclude that the Bayesian algorithm is reliable and practical to enhance the quality of reconstructed images and more suitable method for CT applications.

ACKNOWLEDGEMENTS

We thank M.A Djafari for inspiring discussions and explanation of Bayesian theory during the short stay of Z. Messali in Supelec, France 2011 and for giving her useful feedback on this work.

REFERENCES

1. Q. Zhong, W. Junhao, Y. Pan, "Algebraic Reconstruction Technique in Image Reconstruction with Narrow Fan Beam," IEEE Trans. Medical Imaging., 52(5), pp.1227-1235, 2005.

2. L. Bao-dong, L. Zeng, J. Dong-jiang, "Algebraic Reconstruction Technique Class for Linear Scan CT of Long Object," In. Proc. the 17th World Conference on Nondestructive Testing., Shanghai, China,Oct. 2008.

3. Desai, S. D., Kulkami, L., 2010, A Quantitative Comparative Study of Analytical and Iterative Reconstruction Techniques., International Journal of Processing (IJIP), 4(4), pp. 307-319.

4. Anishinraj, M.M., Venkataraman, B., Vaithiyanathan, V., 2012, A Comparative Study and Experimentation of Simultaneous Algebraic Reconstruction Technique (SART) and Ordered Subsets Expectation Maximization (OSEM).,

5. Eur. J. Sci. Res., 4(68), pp. 584-590.

6. V.H. Tessa, W. Sarah, G. Maggie, B.K. Joost, S. Jan, "The implementation of Iterative Reconstruction Algorithms in MATLAB," Master Thesis, Department of Industrial Sciences and Technology, University College of Antwerp, Belgium, July. 2007.

7. Philippe. P.B., 2002, Analytic and Iterative Reconstruction Algorithms in SPECT., J. Nucl Med. 43(10), 1343-1358.

8. Andersen, A.H., Kak, A.C., 1984, Simultaneous Algebraic Reconstruction Technique (SART) A Superior Implementation of the ART algorithm, Ultrasonics Imaging, 6(1),81–94.

9. A.C. Kak, M. Slaney, "Principles of computerized tomographic imaging," IEEE Press., 1988.

10. D. Raparia, J. Alessi, A. Kponu, "The Algebraic Reconstruction Technique (ART)," Ph.D. dissertation, AGS Department, Brookhaven National Lab, Upton, NY 11973, USA, 1997.

11. H. Carfanton, A.M. Djafari, "A Bayesian Approach For Nonlinear Inverse Scattering Tomographic Imaging," Proceedings of IEEE ICASSP., vol. 4, pp. 2311-2314, USA, 1995.

12. T. Hebert, R. A. Leahy, "Generalized EM Algorithm For 3-D Bayesian Reconstruction From Poisson Data Using Gibbs Priors," IEEE Trans. Med Imaging., vol. 8, pp.194-202, 1989.

13. L. T. Hsiao, A. Rangarajan, G. Gindi, "Joint-Map Reconstruction/ Segmentation for Transmission Tomography Using Mixture-Models as Priors," Proc IEEE Nuclear Science Symposium and Medical Imaging Conference., vol.3,pp. 1689-1693,Toronto, Canada ,Nov. 1998.

14. C. Soussen, A. M. Djafari, "Closed Surface Reconstruction in X-Ray Tomography," Proc IEEE Int Conf Image Processing., vol.1, pp.718–721, Thessaloniki, Greece, Oct. 2001.

Chapter 7

Estimators for Finite Population Variance Using Mean and Variance of Auxiliary Variable

Peeyush Misra, R. Karan Singh

Department of Statistics, Lucknow University, Lucknow, India

ABSTRACT

For estimating finite population variance using information on single auxiliary variable in the form of mean and variance both, the Ratio-Product-Difference (RPD) type estimators are proposed. The generalized cases of these estimators leading to the classes of estimators are also proposed. The bias and mean square error (MSE) of the proposed estimators are found. Theoretical comparisons with the traditional estimator are supported by a numerical example. By this comparison it is shown that the proposed estimators are more efficient than the traditional one.

Keywords: Auxiliary Variable, Taylor's Series Expansion, Bias, Mean Square Error (MSE) and Efficiency

1. INTRODUCTION

In sampling theory, auxiliary information is used widely at both the stages of selection and estimation. At the estimation stage, auxiliary information is used by formulating various types of estimators of different population parameters with a view of getting increased efficiency and are available in plenty in the literature.

Let U = (1, 2, . . . , N) be a finite population of N units with Y being the study variable taking the value Y_i for the unit *i* of U and X being the auxiliary variable taking the value X_i for the unit i of the population, i = 1, 2, . . . , N.

Let $\bar{Y}\left(=\frac{1}{N}\sum_{i=1}^{N}Y_i\right)$ be the population of mean of Y and $\bar{X}\left(=\frac{1}{N}\sum_{i=1}^{N}X_i\right)$ be the population mean of X. Also, let $S_Y^2=\frac{1}{N-1}\sum_{i=1}^{N}\left(Y_i-\bar{Y}\right)^2$, $S_X^2=\frac{1}{N-1}\sum_{i=1}^{N}\left(X_i-\bar{X}\right)^2$ and $\mu_{rs}=\frac{1}{N}\sum_{i=1}^{N}\left(Y_i-\bar{Y}\right)^r\left(X_i-\bar{X}\right)^s$

For a simple random sample of size n drawn from U with the sample observations $y_1, y_2, \ldots, y_n$ on y and $x_1, x_2, \ldots, x_n$ on x, let $\bar{y}=\frac{1}{n}\sum_{i=1}^{n}y_i$ and $\bar{x}=\frac{1}{n}\sum_{i=1}^{n}x_i$ be the sample means of y-values and x-values respectively.

2. THE SUGGESTED ESTIMATORS

We know that the finite population variance σ_Y^2 of the study variable is

$$\sigma_Y^2=\frac{1}{N}\sum_{i=1}^{N}Y_i^2-\bar{Y}^2=\theta-\bar{Y}^2 \text{, where } \theta=\frac{1}{N}\sum_{i=1}^{N}Y_i^2 \qquad (2.1)$$

From (2.1) it is natural to get an estimator of σ_Y^2 if we replace $\theta=\frac{1}{N}\sum_{i=1}^{N}Y_i^2$ and $\bar{Y}^2$ by their some estimators. In particular if $\theta=\frac{1}{N}\sum_{i=1}^{N}Y_i^2$ is estimated by $\hat{\theta}$ and $\bar{Y}^2$ is estimated by $\bar{y}^2\left\{1+\frac{k_1\left(\bar{x}-\bar{X}\right)}{\bar{X}}\right\}\left\{1+\frac{k_2\left(s_x^2-S_x^2\right)}{S_x^2}\right\}$, we get the estimator of σ_Y^2 as follows

$$d_1 = \hat{\theta} - \bar{y}^2 \left\{1 + \frac{k_1(\bar{x} - \bar{X})}{\bar{X}}\right\}\left\{1 + \frac{k_2(s_x^{\ 2} - S_x^{\ 2})}{S_x^{\ 2}}\right\} \quad (2.2)$$

and its generalized estimator as

$$d_{1g} = \hat{\theta} - \bar{y}^2 f(u, v) \quad (2.3)$$

where $u = \frac{\bar{x}}{\bar{X}}$ and $v = \frac{s_x^{\ 2}}{S_x^{\ 2}}$ and f (*u* , *v*) satisfying the validity conditions of Taylor's series expansion is a bounded function of (u , v) such that f(1, 1) = 1.

Also if $\theta = \frac{1}{N}\sum_{i=1}^{N} Y_i^2$ is estimated by $\hat{\theta}\left\{1 + \frac{k_3(\bar{x} - \bar{X})}{\bar{X}}\right\}$ and $\bar{Y}^2$ is estimated by $\bar{y}^2\left\{1 + \frac{k_4(s_x^{\ 2} - S_x^{\ 2})}{S_x^{\ 2}}\right\}$, we get another estimator of $\sigma_Y^{\ 2}$ as follows

$$d_2 = \hat{\theta}\left\{1 + \frac{k_3(\bar{x} - \bar{X})}{\bar{X}}\right\} - \bar{y}^2\left\{1 + \frac{k_4(s_x^{\ 2} - S_x^{\ 2})}{S_x^{\ 2}}\right\} \quad ()2.4$$

and its generalized estimator as

$$d_{2g} = \hat{\theta} f(u) - \bar{y}^2 g(v) \quad (2.5)$$

where $u = \frac{\bar{x}}{\bar{X}}$ and $v = \frac{s_x^{\ 2}}{S_x^{\ 2}}$, f(*u*) and g (*v*) both are bounded functions in u and v respectively such that $f(1)$ = 1 at the point u = 1 and $g(1)$ = 1 at the point v = 1 and both are satisfying the regularity conditions for the validity of Taylor's series expansion having first two derivatives with respect to u and v respectively to be bounded.

3. BIAS AND MEAN SQUARED ERROR OF SUGGESTED ESTIMATORS

(a) Bias and Mean Square Error of Suggested Estimator d_1

Let $\bar{y} = \bar{Y} + e_0$, $\bar{x} = \bar{X} + e_1$, $s_x^{\ 2} = \mu_{02} + e_2$ and $\hat{\theta} = \theta + e_3$,

For simplicity, it is assumed that the population size N is large enough as compared to the sample size n so that finite population correction terms may be ignored. Now

$$E(e_0) = E(e_1) = E(e_2) = E(e_3) = 0 \tag{3.1}$$

$$\left.\begin{aligned}
E(e_0^2) &= \frac{1}{n} S_Y^2 = \frac{\mu_{20}}{n} \\
E(e_1^2) &= \frac{1}{n} S_X^2 = \frac{\mu_{02}}{n} \\
E(e_0 e_1) &= \frac{1}{n} \rho S_Y S_X = \frac{\mu_{11}}{n} \\
E(e_2^2) &= \frac{\mu_{02}^2}{n} (\beta_2 - 1) \\
E(e_0 e_2) &= \frac{\mu_{12}}{n} \\
E(e_1 e_2) &= \frac{\mu_{03}}{n}
\end{aligned}\right\} \tag{3.2}$$

and

$$\left.\begin{aligned}
E(e_3^2) &= \frac{1}{n}(\mu_{40} + 4\bar{Y}\mu_{30} + 4\bar{Y}^2 \mu_{20} - \mu_{20}^2) \\
E(e_0 e_3) &= \frac{1}{n}(\mu_{30} + 2\bar{Y}\mu_{20}) \\
E(e_1 e_3) &= \frac{1}{n}(\mu_{21} + 2\bar{Y}\mu_{11}) \\
E(e_2 e_3) &= \frac{1}{n}(\mu_{22} + 2\bar{Y}\mu_{12} - \mu_{02}\mu_{20})
\end{aligned}\right\} \tag{3.3}$$

Let us consider the proposed estimator d_1 defined in (2.2)

$$d_1 = \hat{\theta} - \bar{y}^2\left\{1 + \frac{k_1(\bar{x} - \bar{X})}{\bar{X}}\right\}\left\{1 + \frac{k_2(s_x^2 - S_x^2)}{S_x^2}\right\}$$

$$= (\theta + e_3) - (\bar{Y} + e_0)^2\left(1 + \frac{k_1 e_1}{\bar{X}}\right)\left(1 + \frac{k_2 e_2}{\mu_{02}}\right)$$

$$= (\theta + e_3) - (\bar{Y}^2 + e_0^2 + 2\bar{Y}e_0)\left(1 + \frac{k_2 e_2}{\mu_{02}} + \frac{k_1 e_1}{\bar{X}} + \frac{k_1 k_2 e_1 e_2}{\bar{X}\mu_{02}}\right)$$

$$= (\theta + e_3) - \bar{Y}^2 - \frac{k_2 \bar{Y}^2 e_2}{\mu_{02}} - \frac{k_1 \bar{Y}^2 e_1}{\bar{X}} - \frac{k_1 k_2 \bar{Y}^2 e_1 e_2}{\bar{X}\mu_{02}} - e_0^2 - 2\bar{Y}e_0 - \frac{2k_2 \bar{Y} e_0 e_2}{\mu_{02}} - \frac{2k_1 \bar{Y} e_0 e_1}{\bar{X}}$$

$$= \left(\frac{1}{N}\sum_{i=1}^{N} Y_i^2 - \bar{Y}^2\right) + e_3 - 2\bar{Y}e_0 - \frac{k_1 \bar{Y}^2 e_1}{\bar{X}} - \frac{k_2 \bar{Y}^2 e_2}{\mu_{02}} - e_0^2 - \frac{2k_1 \bar{Y} e_0 e_1}{\bar{X}} - \frac{2k_2 \bar{Y} e_0 e_2}{\mu_{02}} - \frac{k_1 k_2 \bar{Y}^2 e_1 e_2}{\bar{X}\mu_{02}}$$

Or

$$(d_1 - \sigma_Y^2) = e_3 - 2\bar{Y}e_0 - \frac{k_1 \bar{Y}^2 e_1}{\bar{X}} - \frac{k_2 \bar{Y}^2 e_2}{\mu_{02}} - e_0^2 - \frac{2k_1 \bar{Y} e_0 e_1}{\bar{X}} - \frac{2k_2 \bar{Y} e_0 e_2}{\mu_{02}} - \frac{k_1 k_2 \bar{Y}^2 e_1 e_2}{\bar{X}\mu_{02}}$$

Taking expectation on both sides of (3.4), the bias in $d_1\left(= E(d_1) - \sigma_Y^2\right)$ to the order $\left(\frac{1}{n}\right)$ is given by

Bias

$$(d_1) = E(d_1) - \sigma_Y^2 = E(e_3) - 2\bar{Y}E(e_0) - \frac{k_1 \bar{Y}^2}{\bar{X}}E(e_1) - \frac{k_2 \bar{Y}^2}{\mu_{02}}E(e_2) - E(e_0^2)$$

$$- \frac{2k_1 \bar{Y}}{\bar{X}}E(e_0 e_1) - \frac{2k_2 \bar{Y}}{\mu_{02}}E(e_0 e_2) - \frac{k_1 k_2 \bar{Y}^2}{\bar{X}\mu_{02}}E(e_1 e_2)$$

Using values of the expectations given from (3.1) to (3.3), we have

$$\text{Bias}\,(d_1) = -\frac{\mu_{20}}{n} - \frac{2k_1 \bar{Y}}{\bar{X}}\frac{\mu_{11}}{n} - \frac{2k_2 \bar{Y}}{\mu_{02}}\frac{\mu_{12}}{n} - \frac{k_1 k_2 \bar{Y}^2}{\bar{X}\mu_{02}}\frac{\mu_{03}}{n} \tag{3.5}$$

Now squaring (3.4) on both sides and then taking expectation, the mean square error of $d_1 (= E(d_1) - \sigma_Y^2)^2$ to the first degree of approximation is given by

$$E(d_1-\sigma_Y^2)^2 = E\left(e_3 - 2\bar{Y}e_0 - \frac{k_1\bar{Y}^2 e_1}{\bar{X}} - \frac{k_2\bar{Y}^2 e_2}{\mu_{02}} - e_0^2 - \frac{2k_1\bar{Y}e_0e_1}{\bar{X}} - \frac{2k_2\bar{Y}e_0e_2}{\mu_{02}} - \frac{k_1k_2\bar{Y}^2e_1e_2}{\bar{X}\mu_{02}}\right)^2$$

$$= E(e_3^2) + 4\bar{Y}^2E(e_0^2) + \frac{k_1^2\bar{Y}^4}{\bar{X}^2}E(e_1^2) + \frac{k_2^2\bar{Y}^4}{\mu_{02}^2}E(e_2^2) - 4\bar{Y}E(e_0e_3) - \frac{2k_1\bar{Y}^2}{\bar{X}}E(e_1e_3)$$

$$-\frac{2k_2\bar{Y}^2}{\mu_{02}}E(e_2e_3) + \frac{4k_1\bar{Y}^3}{\bar{X}}E(e_0e_1) + \frac{4k_2\bar{Y}^3}{\mu_{02}}E(e_0e_2) + \frac{2k_1k_2\bar{Y}^4}{\bar{X}\mu_{02}}E(e_1e_2)$$

$$= \frac{1}{n}(\mu_{40} + 4\bar{Y}\mu_{30} + 4\bar{Y}^2\mu_{20} - \mu_{20}^2) + 4\bar{Y}^2\frac{\mu_{20}}{n} - 4\bar{Y}\frac{1}{n}(\mu_{30} + 2\bar{Y}\mu_{20}) + \frac{k_1^2\bar{Y}^4}{\bar{X}^2}\frac{\mu_{02}}{n}$$

$$+\frac{k_2^2\bar{Y}^4}{\mu_{02}^2}\frac{\mu_{02}^2}{n}(\beta_2-1) - \frac{2k_1\bar{Y}^2}{\bar{X}}\frac{1}{n}(\mu_{21} + 2\bar{Y}\mu_{11}) - \frac{2k_2\bar{Y}^2}{\mu_{02}}\frac{1}{n}(\mu_{22} + 2\bar{Y}\mu_{12} - \mu_{02}\mu_{20})$$

$$+\frac{4k_1\bar{Y}^3}{\bar{X}}\frac{\mu_{11}}{n} + \frac{4k_2\bar{Y}^3}{\mu_{02}}\frac{\mu_{12}}{n} + \frac{2k_1k_2\bar{Y}^4}{\bar{X}\mu_{02}}\frac{\mu_{03}}{n}$$

Or

$$\text{MSE}(d_1) = \frac{1}{n}(\mu_{40} - \mu_{20}^2) + \frac{k_1^2\bar{Y}^4}{\bar{X}^2}\frac{\mu_{02}}{n} + \frac{k_2^2\bar{Y}^4}{n}(\beta_2 - 1) - \frac{2k_1\bar{Y}^2}{\bar{X}}\frac{\mu_{21}}{n} - \frac{2k_2\bar{Y}^2}{\mu_{02}}\frac{\mu_{22}}{n}$$

$$+2\bar{Y}^2k_2\frac{\mu_{20}}{n} + \frac{2k_1k_2\bar{Y}^4}{\bar{X}\mu_{02}}\frac{\mu_{03}}{n} \tag{3.6}$$

For minimizing (3.6) in two unknowns k_1 and k_2, the two normal equations after differentiating (3.6) partially with respect to k_1 and k_2 are

$$\frac{2\bar{Y}^4\mu_{02}}{n\bar{X}^2}k_1 - \frac{2\bar{Y}^2\mu_{21}}{n\bar{X}} + \frac{2\bar{Y}^4\mu_{03}}{n\bar{X}\mu_{02}}k_2 = 0 \tag{3.7}$$

$$\frac{2\bar{Y}^4(\beta_2-1)}{n}k_2 + \left(\frac{2\bar{Y}^2\mu_{20}}{n} - \frac{2\bar{Y}^2\mu_{22}}{n\mu_{02}}\right) + \frac{2\bar{Y}^4\mu_{03}}{n\bar{X}\mu_{02}}k_1 = 0 \tag{3.8}$$

Solving (3.7) and (3.8) for k_1 and k_2, we get the minimizing optimum values to be

$$k_1^* = -\frac{\left(\frac{2\bar{Y}^4(\beta_2-1)}{n}\right)\left(\frac{-2\bar{Y}^2\mu_{21}}{n\bar{X}}\right) - \left(\frac{2\bar{Y}^2\mu_{20}}{n} - \frac{2\bar{Y}^2\mu_{22}}{n\mu_{02}}\right)\left(\frac{2\bar{Y}^4\mu_{03}}{n\bar{X}\mu_{02}}\right)}{\left(\frac{2\bar{Y}^2\mu_{02}}{n\bar{X}^2}\right)\left(\frac{2\bar{Y}^4(\beta_2-1)}{n}\right) - \left(\frac{2\bar{Y}^4\mu_{03}}{n\bar{X}\mu_{02}}\right)^2} \tag{3.9}$$

$$k_2^* = -\frac{\left(\frac{2\bar{Y}^4\mu_{02}}{n\bar{X}^2}\right)\left(\frac{2\bar{Y}^2\mu_{20}}{n} - \frac{2\bar{Y}^2\mu_{22}}{n\mu_{02}}\right) - \left(\frac{-2\bar{Y}^2\mu_{21}}{n\bar{X}}\right)\left(\frac{2\bar{Y}^4\mu_{03}}{n\bar{X}\mu_{02}}\right)}{\left(\frac{2\bar{Y}^4\mu_{02}}{n\bar{X}^2}\right)\left(\frac{2\bar{Y}^4(\beta_2-1)}{n}\right) - \left(\frac{2\bar{Y}^4\mu_{03}}{n\bar{X}\mu_{02}}\right)^2} \tag{3.10}$$

which when substituted in (3.6) gives the minimum value of mean square error of the estimator d_1 as

$$\text{MSE}(d_1)\text{min} = \frac{1}{n}\left(\mu_{40} - \mu_{20}^2\right) - \frac{\left\{\frac{2\mu_{21}\mu_{03}}{n\mu_{02}}\left(\mu_{20} - \frac{\mu_{22}}{\mu_{02}}\right) - \frac{(\beta_2-1)\mu_{21}^2}{n} - \frac{\mu_{02}}{n}\left(\mu_{20} - \frac{\mu_{22}}{\mu_{02}}\right)^2\right\}}{\left\{(\beta_2-1)\mu_{02} - \frac{\mu_{03}^2}{\mu_{02}^2}\right\}} \tag{3.11}$$

(b) Bias and Mean Square Error of Suggested Estimator d_{1g}

For f_1 and f_2 being the first order partial derivatives of $f(u,v)$ with respect to u and v respectively at the point (1,1), that is

$$\left(\frac{\partial}{\partial u} f(u,v)\right)_{(1,1)} = f_1 \text{ and } \left(\frac{\partial}{\partial v} f(u,v)\right)_{(1,1)} = f_2,$$

expanding $f(u,v)$ in (2.3) in third order Taylor's series about point (1, 1), we have

$$d_{1g} = \hat{\theta} - \bar{y}^2\Big[f(1,1) + (u-1)f_1 + (v-1)f_2 + \frac{1}{2!}\{(u-1)^2 f_{11} + (v-1)^2 f_{22} + 2(u-1)(v-1)f_{12}\}$$
$$+ \frac{1}{3!}\left\{(u-1)\frac{\partial}{\partial u} + (v-1)\frac{\partial}{\partial v}\right\}^3 f(u^*,v^*)\Big]$$

where f_1 and f_2 are already defined; f_{11}, f_{22} and f_{12} are the second order partial derivatives given by

$$f_{11} = \left(\frac{\partial^2}{\partial u^2} f(u,v)\right)_{(1,1)}, f_{22} = \left(\frac{\partial^2}{\partial v^2} f(u,v)\right)_{(1,1)}, f_{12} = \left(\frac{\partial^2}{\partial u \partial v} f(u,v)\right)_{(1,1)}$$

and $u^* = 1 + h(u-1)$, $v^* = 1 + h(v-1)$, $0 < h < 1$.

or

$$d_{1g} = (\theta + e_3) - (\bar{Y} + e_0)^2 \left[1 + \frac{e_1}{\bar{X}} f_1 + \frac{e_2}{\mu_{02}} f_2 + \frac{1}{2!}\left\{ \frac{e_1^2}{\bar{X}^2} f_{11} + \frac{e_2^2}{\mu_{02}^2} f_{22} + 2\frac{e_1 e_2}{\bar{X}\mu_{02}} f_{12} \right\} \right.$$

$$\left. + \frac{1}{3!}\left\{ \frac{e_1}{\bar{X}}\frac{\partial}{\partial u} + \frac{e_2}{\mu_{02}}\frac{\partial}{\partial v} \right\}^3 f(u^*, v^*) \right]$$

$$= (\theta + e_3) - \bar{Y}^2 - \frac{\bar{Y}^2 e_1}{\bar{X}} f_1 - \frac{\bar{Y}^2 e_2}{\mu_{02}} f_2 - \frac{\bar{Y}^2}{2}\left(\frac{e_1^2}{\bar{X}^2} f_{11} + \frac{e_2^2}{\mu_{02}^2} f_{22} + 2\frac{e_1 e_2}{\bar{X}\mu_{02}} f_{12} \right)$$

$$- e_0^2 - 2\bar{Y}e_0 - \frac{2\bar{Y}e_0 e_1}{\bar{X}} f_1 - \frac{2\bar{Y}e_0 e_2}{\mu_{02}} f_2$$

$$= \left(\frac{1}{N}\sum_{i=1}^{N} Y_i^2 - \bar{Y}^2 \right) + e_3 - 2\bar{Y}e_0 - \frac{\bar{Y}^2 e_1}{\bar{X}} f_1 - \frac{\bar{Y}^2 e_2}{\mu_{02}} f_2 - e_0^2 - \frac{\bar{Y}^2 e_1^2}{2\bar{X}^2} f_{11} - \frac{\bar{Y}^2 e_2^2}{2\mu_{02}} f_{22}$$

$$- \frac{2\bar{Y}e_0 e_1}{\bar{X}} f_1 - \frac{2\bar{Y}e_0 e_2}{\mu_{02}} f_2 - \frac{\bar{Y}^2 e_1 e_2}{\bar{X}\mu_{02}} f_{12}$$

$$d_{1g} - \sigma_Y^2 = e_3 - 2\bar{Y}e_0 - \frac{\bar{Y}^2 e_1}{\bar{X}} f_1 - \frac{\bar{Y}^2 e_2}{\mu_{02}} f_2 - e_0^2 - \frac{\bar{Y}^2 e_1^2}{2\bar{X}^2} f_{11} - \frac{\bar{Y}^2 e_2^2}{2\mu_{02}^2} f_{22}$$

$$- \frac{2\bar{Y}e_0 e_1}{\bar{X}} f_1 - \frac{2\bar{Y}e_0 e_2}{\mu_{02}} f_2 - \frac{\bar{Y}^2 e_1 e_2}{\bar{X}\mu_{02}} f_{12}$$

Taking expectation on both sides of (3.12) and using values of the expectations given from (3.1) to (3.3), the bias in d_{1g} $\left(= E(d_{1g}) - \sigma_Y^2\right)$ to the order $\left(\frac{1}{n}\right)$ is given by

$$\text{Bias}(d_{1g}) = E(d_{1g}) - \sigma_Y^2$$

$$= -\frac{\mu_{20}}{n} - \frac{\bar{Y}^2}{2\bar{X}^2} f_{11}\frac{\mu_{02}}{n} - \frac{\bar{Y}^2}{2\mu_{02}^2} f_{22}\frac{\mu_{02}^2}{n}(\beta_2 - 1) - \frac{2\bar{Y}}{\bar{X}} f_1 \frac{\mu_{11}}{n} - \frac{2\bar{Y}}{\mu_{02}} f_2 \frac{\mu_{12}}{n} - \frac{\bar{Y}^2}{\bar{X}\mu_{02}} f_{12}\frac{\mu_{03}}{n}$$

$$\text{Bias}(d_{1g}) = -\frac{\mu_{20}}{n} - \frac{2\bar{Y}}{\bar{X}}\frac{\mu_{11}}{n} f_1 - \frac{2\bar{Y}}{\mu_{02}}\frac{\mu_{12}}{n} f_2 - \frac{\bar{Y}^2}{2\bar{X}^2}\frac{\mu_{02}}{n} f_{11} - \frac{\bar{Y}^2}{2n} f_{22}(\beta_2 - 1) - \frac{\bar{Y}^2}{\bar{X}\mu_{02}}\frac{\mu_{03}}{n} f_{12}$$

(3.13)

Now squaring (3.12) on both sides and then taking expectation, the mean square error of d_{1g} to the first degree of approximation is given by

$$E(d_{1g}-\sigma_Y^2)^2 = E\left(e_3 - 2\bar{Y}e_0 - \frac{\bar{Y}^2 e_1}{\bar{X}} f_1 - \frac{\bar{Y}^2 e_2}{\mu_{02}} f_2\right)^2$$

$$= E\left(e_3^2\right) + 4\bar{Y}^2 E\left(e_0^2\right) + \frac{\bar{Y}^4}{\bar{X}^2} f_1^2 E\left(e_1^2\right) + \frac{\bar{Y}^4}{\mu_{02}^2} f_2^2 E\left(e_2^2\right) - 4\bar{Y}E(e_0 e_3) - \frac{2\bar{Y}^2}{\bar{X}} f_1 E(e_1 e_3)$$

$$-\frac{2\bar{Y}^2}{\mu_{02}} f_2 E\left(e_2 e_3\right) + \frac{4\bar{Y}^3}{\bar{X}} f_1 E\left(e_0 e_1\right) + \frac{4\bar{Y}^3}{\mu_{02}} f_2 E\left(e_0 e_2\right) + \frac{2\bar{Y}^4}{\bar{X}\mu_{02}} f_1 f_2 E\left(e_1 e_2\right)$$

$$= \frac{1}{n}\left(\mu_{40} + 4\bar{Y}\mu_{30} + 4\bar{Y}^2\mu_{20} - \mu_{20}^2\right) + 4\bar{Y}^2\frac{\mu_{20}}{n} - 4\bar{Y}\frac{1}{n}\left(\mu_{30} + 2\bar{Y}\mu_{20}\right) + \frac{\bar{Y}^4}{\bar{X}^2} f_1^2 \frac{\mu_{02}}{n}$$

$$+\frac{\bar{Y}^4}{\mu_{02}^2} f_2^2 \frac{\mu_{02}^2(\beta_2 - 1)}{n} - \frac{2\bar{Y}^2}{\bar{X}} f_1 \frac{1}{n}(\mu_{21} + 2\bar{Y}\mu_{11}) - \frac{2\bar{Y}^2}{\mu_{02}} f_2 \frac{1}{n}\left(\mu_{22} + 2\bar{Y}\mu_{12} - \mu_{02}\mu_{20}\right)$$

$$+\frac{4\bar{Y}^3}{\bar{X}} f_1 \frac{\mu_{11}}{n} + \frac{4\bar{Y}^3}{\mu_{02}} f_2 \frac{\mu_{12}}{n} + \frac{2\bar{Y}^4}{\bar{X}\mu_{02}} f_1 f_2 \frac{\mu_{03}}{n}$$

$$\text{MSE}\,(d_{1g}) = \frac{1}{n}\left(\mu_{40} - \mu_{20}^2\right) + \frac{\bar{Y}^4}{\bar{X}^2} f_1^2 \frac{\mu_{02}}{n} + \frac{\bar{Y}^4(\beta_2 - 1)}{n} f_2^2 - \frac{2\bar{Y}^2}{\bar{X}} f_1 \frac{\mu_{21}}{n} - \frac{2\bar{Y}^2}{\mu_{02}} f_2 \frac{\mu_{22}}{n}$$

$$+ 2\bar{Y}^2 f_2 \frac{\mu_{20}}{n} + \frac{2\bar{Y}^4}{\bar{X}\mu_{02}} f_1 f_2 \frac{\mu_{03}}{n}$$

For minimizing (3.14) in two unknowns f_1 and f_2, the two normal equations after differentiating (3.14) partially with respect to f_1 and f_2 are

$$\frac{2\bar{Y}^4\mu_{02}}{n\bar{X}^2} f_1 - \frac{2\bar{Y}^2\mu_{21}}{n\bar{X}} + \frac{2\bar{Y}^4\mu_{03}}{n\bar{X}\mu_{02}} f_2 = 0 \qquad (3.15)$$

$$\frac{2\bar{Y}^4(\beta_2 - 1)}{n} f_2 + \left(\frac{2\bar{Y}^2\mu_{20}}{n} - \frac{2\bar{Y}^2\mu_{22}}{n\mu_{02}}\right) + \frac{2\bar{Y}^4\mu_{03}}{n\bar{X}\mu_{02}} f_1 = 0 \qquad (3.16)$$

Solving (3.15) and (3.16) for f_1 and f_2, we get the minimizing optimum values to be

$$f_1^* = -\frac{\left(\frac{2\bar{Y}^4(\beta_2 - 1)}{n}\right)\left(\frac{-2\bar{Y}^2\mu_{21}}{n\bar{X}}\right) - \left(\frac{2\bar{Y}^2\mu_{20}}{n} - \frac{2\bar{Y}^2\mu_{22}}{n\mu_{02}}\right)\left(\frac{2\bar{Y}^4\mu_{03}}{n\bar{X}\mu_{02}}\right)}{\left(\frac{2\bar{Y}^2\mu_{02}}{n\bar{X}^2}\right)\left(\frac{2\bar{Y}^4(\beta_2 - 1)}{n}\right) - \left(\frac{2\bar{Y}^4\mu_{03}}{n\bar{X}\mu_{02}}\right)^2} \qquad (3.17)$$

$$f_2^* = -\frac{\left(\frac{2\bar{Y}^4\mu_{02}}{n\bar{X}^2}\right)\left(\frac{2\bar{Y}^2\mu_{20}}{n} - \frac{2\bar{Y}^2\mu_{22}}{n\mu_{02}}\right) - \left(\frac{-2\bar{Y}^2\mu_{21}}{n\bar{X}}\right)\left(\frac{2\bar{Y}^4\mu_{03}}{n\bar{X}\mu_{02}}\right)}{\left(\frac{2\bar{Y}^4\mu_{02}}{n\bar{X}^2}\right)\left(\frac{2\bar{Y}^4(\beta_2 - 1)}{n}\right) - \left(\frac{2\bar{Y}^4\mu_{03}}{n\bar{X}\mu_{02}}\right)^2} \qquad (3.18)$$

which when substituted in (3.14) gives the minimum value of mean square error of the estimator d_{1g} as

$$\text{MSE}(d_{1g})\min = \frac{1}{n}\left(\mu_{40} - \mu_{20}^{2}\right) - \frac{\left\{\frac{2\mu_{21}\mu_{03}}{n\mu_{02}}\left(\mu_{20} - \frac{\mu_{22}}{\mu_{02}}\right) - \frac{(\beta_2 - 1)\mu_{21}^{2}}{n} - \frac{\mu_{02}}{n}\left(\mu_{20} - \frac{\mu_{22}}{\mu_{02}}\right)^{2}\right\}}{\left\{(\beta_2 - 1)\mu_{02} - \frac{\mu_{03}^{2}}{\mu_{02}^{2}}\right\}} \tag{3.19}$$

(c) Bias and Mean Square Error of Suggested Estimator d_2

Let us consider the proposed estimator d_2 defined in (2.4)

$$d_2 = \hat{\theta}\left\{1 + \frac{k_3(\bar{x} - \bar{X})}{\bar{X}}\right\} - \bar{y}^{2}\left\{1 + \frac{k_4(s_x^{2} - S_x^{2})}{S_x^{2}}\right\}$$

$$= \hat{\theta}\left\{1 + \frac{k_3(\bar{X} + e_1 - \bar{X})}{\bar{X}}\right\} - \bar{y}^{2}\left\{1 + \frac{k_4(S_x^{2} + e_2 - S_x^{2})}{S_x^{2}}\right\}$$

$$= (\theta + e_3)\left\{1 + \frac{k_3 e_1}{\bar{X}}\right\} - (\bar{Y}^{2} + e_0^{2} + 2\bar{Y}e_0)\left\{1 + \frac{k_4 e_2}{\mu_{02}}\right\}$$

$$= \theta + \frac{\theta\, k_3 e_1}{\bar{X}} + e_3 + \frac{k_3 e_1 e_3}{\bar{X}} - \bar{Y}^{2} - \frac{\bar{Y}^{2} k_4 e_2}{\mu_{02}} - e_0^{2} - 2\bar{Y}e_0 - \frac{2\bar{Y}k_4 e_0 e_2}{\mu_{02}}$$

$$= \left(\frac{1}{N}\sum_{i=1}^{N} Y_i^{2} - \bar{Y}^{2}\right) + e_3 - 2\bar{Y}e_0 + \frac{\theta\, k_3 e_1}{\bar{X}} - \frac{\bar{Y}^{2} k_4 e_2}{\mu_{02}} - e_0^{2} + \frac{k_3 e_1 e_3}{\bar{X}} - \frac{2\bar{Y}k_4 e_0 e_2}{\mu_{02}}$$

or

$$d_2 - \sigma_Y^{2} = e_3 - 2\bar{Y}e_0 + \frac{\theta\, k_3 e_1}{\bar{X}} - \frac{\bar{Y}^{2} k_4 e_2}{\mu_{02}} - e_0^{2} + \frac{k_3 e_1 e_3}{\bar{X}} - \frac{2\bar{Y}k_4 e_0 e_2}{\mu_{02}} \tag{3.20}$$

Taking expectation on both sides of (3.20), the bias in $d_2\ (= E(d_2) - \sigma_Y^{2})$ up to terms of order $\left(\frac{1}{n}\right)$ is given by

$$\text{Bias}(d_2) = E(e_3) - 2\bar{Y}E(e_0) + \frac{\theta\, k_3 E(e_1)}{\bar{X}} - \frac{\bar{Y}^{2} k_4 E(e_2)}{\mu_{02}} - E(e_0^{2}) + \frac{k_3 E(e_1 e_3)}{\bar{X}} - \frac{2\bar{Y}k_4 E(e_0 e_2)}{\mu_{02}}$$

Using values of the expectations given from (3.1) to (3.3), we have

$$\text{Bias}(d_2) = E(d_2) - \sigma_Y^{2} = \frac{k_3}{\bar{X}}\frac{1}{n}\left(\mu_{21} + 2\bar{Y}\mu_{11}\right) - \frac{2\bar{Y}k_4}{\mu_{02}}\frac{\mu_{12}}{n} - \frac{\mu_{20}}{n} \tag{3.21}$$

Now squaring (3.20) on both sides and then taking expectation, the mean square error to the first degree of approximation is

$$E(d_2-\sigma_{\bar{Y}}^2)^2 = E\left(e_3-2\bar{Y}e_0+\frac{\theta\, k_3 e_1}{\bar{X}}-\frac{\bar{Y}^2 k_4 e_2}{\mu_{02}}\right)^2$$

$$= E\left(e_3^{\,2}\right)+4\bar{Y}^2E\left(e_0^{\,2}\right)+\frac{\theta^2 k_3^{\,2}}{\bar{X}^2}E\left(e_1^{\,2}\right)+\frac{\bar{Y}^4 k_4^{\,2}}{\mu_{02}^{\,2}}E\left(e_2^{\,2}\right)-4\bar{Y}E\left(e_0e_3\right)+\frac{2\theta\, k_3}{\bar{X}}E\left(e_1e_3\right)$$

$$-\frac{2\bar{Y}^2k_4}{\mu_{02}}E\left(e_2e_3\right)-\frac{4\bar{Y}\theta\, k_3}{\bar{X}}E\left(e_0e_1\right)+\frac{4\bar{Y}^3k_4}{\mu_{02}}E\left(e_0e_2\right)-\frac{2\theta\,\bar{Y}^2k_3k_4}{\bar{X}\mu_{02}}E\left(e_1e_2\right)$$

$$=\frac{1}{n}(\mu_{40}+4\bar{Y}\mu_{30}+4\bar{Y}^2\mu_{20}-\mu_{20}^{\,2})+4\bar{Y}^2\frac{\mu_{20}}{n}-4\bar{Y}\frac{1}{n}(\mu_{30}+2\bar{Y}\mu_{20})+\frac{\theta^2}{\bar{X}^2}k_3^{\,2}\frac{\mu_{02}}{n}$$

$$+\frac{\bar{Y}^4k_4^{\,2}}{\mu_{02}^{\,2}}\frac{\mu_{02}^{\,2}}{n}(\beta_2-1)+\frac{2\theta\, k_3}{\bar{X}}\frac{1}{n}\left(\mu_{21}+2\bar{Y}\mu_{11}\right)-\frac{2\bar{Y}^2k_4}{\mu_{02}}\frac{1}{n}(\mu_{22}+2\bar{Y}\mu_{12}-\mu_{02}\mu_{20})$$

$$-\frac{4\bar{Y}\theta\, k_3}{\bar{X}}\frac{\mu_{11}}{n}+\frac{4\bar{Y}^3k_4}{\mu_{02}}\frac{\mu_{12}}{n}-\frac{2\theta\,\bar{Y}^2k_3k_4}{\bar{X}\mu_{02}}\frac{\mu_{03}}{n}$$

Or

$$\text{MSE}(d_2)=\frac{1}{n}\left(\mu_{40}-\mu_{20}^{\,2}\right)+\frac{\theta^2k_3^{\,2}}{\bar{X}^2}\frac{\mu_{02}}{n}+\frac{\bar{Y}^4k_4^{\,2}}{n}(\beta_2-1)+\frac{2\theta\, k_3}{\bar{X}}\frac{\mu_{21}}{n}-\frac{2\bar{Y}^2k_4}{\mu_{02}}\frac{\mu_{22}}{n}$$

$$+2\,\bar{Y}^2k_4\frac{\mu_{20}}{n}-\frac{2\theta\,\bar{Y}^2k_3k_4}{\bar{X}\mu_{02}}\frac{\mu_{03}}{n}$$

For minimizing (3.22) in two unknowns k_3 and k_4, the two normal equations after differentiating (3.22) partially with respect to k_3 and k_4 are

$$\frac{2\theta^2}{n\bar{X}^2}\mu_{02}k_3+\frac{2\theta}{n\bar{X}}\mu_{21}-\frac{2\theta\bar{Y}^2\mu_{03}k_4}{n\bar{X}\mu_{02}}=0 \quad (3.23)$$

$$\frac{2\bar{Y}^4}{n}(\beta_2-1)k_4+\left(\frac{2\bar{Y}^2\mu_{20}}{n}-\frac{2\bar{Y}^2}{\mu_{02}}\frac{\mu_{22}}{n}\right)-\frac{2\theta\bar{Y}^2\mu_{03}}{n\bar{X}\mu_{02}}k_3=0 \quad (3.24)$$

Solving (3.23) and (3.24) for k_3 and k_4, we get the minimizing optimum values to be

$$k_3* = -\frac{\left(\frac{2\bar{Y}^4(\beta_2-1)}{n}\right)\left(\frac{2\theta\mu_{21}}{n\bar{X}}\right)-\left(\frac{2\bar{Y}^2\mu_{20}}{n}-\frac{2\bar{Y}^2}{\mu_{02}}\frac{\mu_{22}}{n}\right)\left(-\frac{2\theta\bar{Y}^2\mu_{03}}{n\bar{X}\mu_{02}}\right)}{\left(\frac{2\theta^2\mu_{02}}{n\bar{X}^2}\right)\left(\frac{2\bar{Y}^4(\beta_2-1)}{n}\right)-\left(\frac{-2\theta\bar{Y}^2\mu_{03}}{n\bar{X}\mu_{02}}\right)^2} \tag{3.25}$$

And

$$k_4* = -\frac{\left(\frac{2\theta^2\mu_{02}}{n\bar{X}^2}\right)\left(\frac{2\bar{Y}^2\mu_{20}}{n}-\frac{2\bar{Y}^2\mu_{22}}{n\mu_{02}}\right)-\left(\frac{2\theta\mu_{21}}{n\bar{X}}\right)\left(\frac{-2\theta\bar{Y}^2\mu_{03}}{n\bar{X}\mu_{02}}\right)}{\left(\frac{2\theta^2\mu_{02}}{n\bar{X}^2}\right)\left(\frac{2\bar{Y}^4(\beta_2-1)}{n}\right)-\left(\frac{-2\theta\bar{Y}^2\mu_{03}}{n\bar{X}\mu_{02}}\right)^2}$$

which when substituted in (3.22) gives the minimum value of mean square error of the estimator to be

$$\mathrm{MSE}(d_2)_{\min} = \frac{1}{n}\left(\mu_{40}-\mu_{20}{}^2\right) - \frac{\left\{\left(\frac{2\mu_{21}\mu_{03}}{n\mu_{02}}\right)\left(\mu_{20}-\frac{\mu_{22}}{\mu_{02}}\right)+\frac{\mu_{02}}{n}\left(\mu_{20}-\frac{\mu_{22}}{\mu_{02}}\right)^2+\frac{(\beta_2-1)}{n}\mu_{21}{}^2\right\}}{\left\{\mu_{02}(\beta_2-1)-\left(\frac{\mu_{03}}{\mu_{02}}\right)^2\right\}} \tag{3.26}$$

(d) Bias and Mean Square Error of Suggested Estimator d_{2g}

For $f'(1)$, $f''(1)$ and $f'''(1)$ to be first, second and third order derivatives of $f(u)$ at the point u = 1, u* = 1 + h_1 (u-1), 0 < h_1 < 1 , also $g'(1)$, $g''(1)$ and $g'''(1)$ to be first, second and third order derivatives of $g(v)$ at the point v = 1 and v* = 1 + h_2 (v-1), 0 < h_2 < 1, expanding $f(u)$ and $g(v)$ in d_{2g} in third order Taylor's series, we have

$$d_{2g} = \hat{\theta}\left\{ f(1) + (u-1)f'(1) + \frac{(u-1)^2}{2!} f''(1) + \frac{(u-1)^3}{3!} f'''(u^*) \right\}$$
$$- \bar{y}^2 \left\{ g(1) + (v-1)g'(1) + \frac{(v-1)^2}{2!} g''(1) + \frac{(v-1)^3}{3!} g'''(v^*) \right\}$$
$$= \hat{\theta}\left\{ 1 + \frac{e_1}{\bar{X}} f'(1) + \frac{1}{2!}\frac{e_1^2}{\bar{X}^2} f''(1) + \frac{1}{3!}\frac{e_1^3}{\bar{X}^3} f'''(u^*) \right\} - \bar{y}^2 \left\{ 1 + \frac{e_2}{\mu_{02}} g'(1) + \frac{1}{2!}\frac{e_2^2}{\mu_{02}^2} g''(1) + \frac{1}{3!}\frac{e_2^3}{\mu_{02}^3} g'''(v^*) \right\}$$
$$= (\theta + e_3)\left\{ 1 + \frac{e_1}{\bar{X}} f'(1) + \frac{1}{2}\frac{e_1^2}{\bar{X}^2} f''(1) \right\} - \left(\bar{Y}^2 + e_0^2 + 2\bar{Y}e_0 \right) \cdot \left\{ 1 + \frac{e_2}{\mu_{02}} g'(1) + \frac{1}{2}\frac{e_2^2}{\mu_{02}^2} g''(1) \right\}$$
$$= \theta + \frac{\theta e_1}{\bar{X}} f'(1) + \frac{\theta e_1^2}{2\bar{X}^2} f''(1) + e_3 + \frac{e_1 e_3}{\bar{X}} f'(1) - \bar{Y}^2 - \frac{\bar{Y}^2 e_2}{\mu_{02}} g'(1) - \frac{\bar{Y}^2}{2}\frac{e_2^2}{\mu_{02}^2} g''(1) - e_0^2 - 2\bar{Y}e_0 - \frac{2\bar{Y}e_0 e_2}{\mu_{02}} g'(1)$$
$$= \left(\frac{1}{N}\sum_{i=1}^{N} Y_i^2 - \bar{Y}^2 \right) + e_3 - 2\bar{Y}e_0 + \frac{\theta e_1}{\bar{X}} f'(1) - \frac{\bar{Y}^2 e_2}{\mu_{02}} g'(1) - e_0^2 + \frac{\theta e_1^2}{2\bar{X}^2} f''(1)$$
$$- \frac{\bar{Y}^2 e_2^2}{2\mu_{02}^2} g''(1) + \frac{e_1 e_3}{\bar{X}} f'(1) - \frac{2\bar{Y}e_0 e_2}{\mu_{02}} g'(1) \tag{3.27}$$

Or

$$d_{2g} - \sigma_Y^2 = e_3 - 2\bar{Y}e_0 + \frac{\theta e_1}{\bar{X}} f'(1) - \frac{\bar{Y}^2 e_2}{\mu_{02}} g'(1) - e_0^2 + \frac{\theta e_1^2}{2\bar{X}^2} f''(1)$$
$$- \frac{\bar{Y}^2 e_2^2}{2\mu_{02}^2} g''(1) + \frac{e_1 e_3}{\bar{X}} f'(1) - \frac{2\bar{Y}e_0 e_2}{\mu_{02}} g'(1) \tag{3.28}$$

Taking expectation on both sides of (3.28) and using values of the expectations given from (3.1) to (3.3), the bias in $d_{2g}\ (= E(d_{2g}) - \sigma_Y^2)$ to the order $\left(\frac{1}{n}\right)$ is given by

$$\text{Bias}(d_{2g}) = -\frac{\mu_{20}}{n} + \frac{\theta^2}{2\bar{X}^2} f''(1)\frac{\mu_{02}}{n} - \frac{\bar{Y}^2}{2\mu_{02}^2} g''(1)\frac{\mu_{04}}{n} + \frac{1}{\bar{X}} f'(1)\frac{1}{n}\left(\mu_{21} + 2\bar{Y}\mu_{11}\right) - \frac{2\bar{Y}}{\mu_{02}} g'(1)\frac{\mu_{12}}{n}$$
$$= \frac{\theta^2}{2\bar{X}^2} f''(1)\frac{\mu_{02}}{n} - \frac{\bar{Y}^2}{2\mu_{02}^2} g''(1)\frac{\mu_{04}}{n} + \frac{1}{\bar{X}} f'(1)\frac{1}{n}\left(\mu_{21} + 2\bar{Y}\mu_{11}\right) - \frac{2\bar{Y}}{\mu_{02}} g'(1)\frac{\mu_{12}}{n} - \frac{\mu_{20}}{n}$$

Now squaring (3.28) on both sides and then taking expectation, the mean square error of d_{2g} to the first degree of approximation is given by

$$E(d_{2g}-\sigma_Y^2)^2 = E\left(e_3 - 2\bar{Y}e_0 + \frac{\theta e_1}{\bar{X}} f'(1) - \frac{\bar{Y}^2 e_2}{\mu_{02}} g'(1)\right)^2$$

$$= E(e_3^2) + 4\bar{Y}^2 E(e_0^2) + \frac{\theta^2}{\bar{X}^2}\{f'(1)\}^2 E(e_1^2) + \frac{\bar{Y}^4}{\mu_{02}^2}\{g'(1)\}^2 E(e_2^2) - 4\bar{Y}E(e_0 e_3) + \frac{2\theta}{\bar{X}} f'(1)E(e_1 e_3)$$

$$-\frac{2\bar{Y}^2}{\mu_{02}} g'(1)E(e_2 e_3) - \frac{4\bar{Y}\theta}{\bar{X}} f'(1)E(e_0 e_1) + \frac{4\bar{Y}^3}{\mu_{02}} g'(1)E(e_0 e_2) - \frac{2\bar{Y}^2\theta}{\bar{X}\mu_{02}} f'(1)g'(1)E(e_1 e_2)$$

Using values of the expectations given from (3.1) to (3.3), we have

$$\text{MSE}(d_{2g}) = E(d_{2g} - \sigma_Y^2)^2$$

$$= \frac{1}{n}(\mu_{40} + 4\bar{Y}\mu_{30} + 4\bar{Y}^2\mu_{20} - \mu_{20}^2) + 4\bar{Y}^2\frac{\mu_{20}}{n}$$

$$-4\bar{Y}\frac{1}{n}(\mu_{30} + 2\bar{Y}\mu_{20}) + \frac{\theta^2}{\bar{X}^2}\{f'(1)\}^2\frac{\mu_{02}}{n} + \frac{\bar{Y}^4}{\mu_{02}^2}\{g'(1)\}^2\frac{\mu_{02}^2}{n}(\beta_2 - 1) + \frac{2\theta}{\bar{X}} f'(1)\frac{1}{n}(\mu_{21} + 2\bar{Y}\mu_{11})$$

$$-\frac{2\bar{Y}^2}{\mu_{02}} g'(1)\frac{1}{n}(\mu_{22} + 2\bar{Y}\mu_{12} - \mu_{02}\mu_{20}) - \frac{4\bar{Y}\theta}{\bar{X}} f'(1)\frac{\mu_{11}}{n} + \frac{4\bar{Y}^3}{\mu_{02}} g'(1)\frac{\mu_{12}}{n} - \frac{2\bar{Y}^2\theta}{\bar{X}\mu_{02}} f'(1)g'(1)\frac{\mu_{03}}{n}$$

Or

$$\text{MSE}(d_{2g}) = E(d_{2g} - \sigma_Y^2)^2$$

$$= \frac{1}{n}(\mu_{40} - \mu_{20}^2) + \frac{\theta^2}{\bar{X}^2}\{f'(1)\}^2\frac{\mu_{02}}{n} + \frac{\bar{Y}^4(\beta_2 - 1)}{n}\{g'(1)\}^2 + \frac{2\theta}{\bar{X}} f'(1)\frac{\mu_{21}}{n} - \frac{2\bar{Y}^2}{\mu_{02}}\frac{\mu_{22}}{n} g'(1)$$

$$+ 2\bar{Y}^2\frac{\mu_{20}}{n} g'(1) - \frac{2\bar{Y}^2\theta}{\bar{X}\mu_{02}}\frac{\mu_{03}}{n} f'(1)g'(1)$$

For minimizing (3.30) in two unknowns $f'(1)$ and $g'(1)$, the two normal equations after differentiating (3.30) partially with respect to $f'(1)$ and $g'(1)$ are

$$\frac{2\theta^2\mu_{02}}{n\bar{X}^2} f'(1) + \frac{2\theta\,\mu_{21}}{n\bar{X}} - \frac{2\bar{Y}^2\theta\,\mu_{03}}{n\bar{X}\mu_{02}} g'(1) = 0 \tag{3.31}$$

And

$$\frac{2\bar{Y}^4(\beta_2 - 1)}{n} g'(1) + \left(\frac{2\bar{Y}^2\mu_{20}}{n} - \frac{2\bar{Y}^2\mu_{22}}{n\mu_{02}}\right) - \frac{2\bar{Y}^2\theta\mu_{03}}{n\bar{X}\mu_{02}} f'(1) = 0 \tag{3.32}$$

Solving (3.31) and (3.32) for $f'(1)$ and $g'(1)$, we get the minimizing optimum values to be

$$f'(1)^* = -\frac{\left(\frac{2\overline{Y}^4(\beta_2-1)}{n}\right)\left(\frac{2\theta\mu_{21}}{n\overline{X}}\right)-\left(2\overline{Y}^2\frac{\mu_{20}}{n}-\frac{2\overline{Y}^2}{\mu_{02}}\frac{\mu_{22}}{n}\right)\left(-\frac{2\overline{Y}^2\theta\mu_{03}}{n\overline{X}\mu_{02}}\right)}{\left(\frac{2\theta^2\mu_{02}}{n\overline{X}^2}\right)\left(\frac{2\overline{Y}^4(\beta_2-1)}{n}\right)-\left(-\frac{2\overline{Y}^2\theta\ \mu_{03}}{n\overline{X}\mu_{02}}\right)^2} \tag{3.33}$$

$$g'(1)^* = -\frac{\left(\frac{2\theta^2\mu_{02}}{n\overline{X}^2}\right)\left(\frac{2\overline{Y}^2\mu_{20}}{n}-\frac{2\overline{Y}^2\mu_{22}}{n\mu_{02}}\right)-\left(\frac{2\theta\ \mu_{21}}{n\overline{X}}\right)\left(-\frac{2\overline{Y}^2\theta\ \mu_{03}}{n\overline{X}\mu_{02}}\right)}{\left(\frac{2\theta^2\mu_{02}}{n\overline{X}^2}\right)\left(\frac{2\overline{Y}^4(\beta_2-1)}{n}\right)-\left(-\frac{2\overline{Y}^2\theta\ \mu_{03}}{n\overline{X}\mu_{02}}\right)^2} \tag{3.34}$$

which when substituted in (3.30) gives the minimum value of mean square error as

$$\mathrm{MSE}(d_{2g})_{\min} = \frac{1}{n}\left(\mu_{40}-\mu_{20}^2\right)-\frac{\left\{\left(\frac{2\mu_{21}\mu_{03}}{n\mu_{02}}\right)\left(\mu_{20}-\frac{\mu_{22}}{\mu_{02}}\right)+\frac{\mu_{02}}{n}\left(\mu_{20}-\frac{\mu_{22}}{\mu_{02}}\right)^2+\frac{(\beta_2-1)}{n}\mu_{21}^2\right\}}{\left\{\mu_{02}(\beta_2-1)-\left(\frac{\mu_{03}}{\mu_{02}}\right)^2\right\}} \tag{3.35}$$

4. EFFICIENCY COMPARISON WITH THE TRADITIONAL ESTIMATOR

As we know that the mean square error of usual conventional unbiased estimator $s^2 = \frac{1}{n-1}\sum_{i=1}^{n}(y_i-\bar{y})^2$ of population variance σ_Y^2 is $\frac{1}{n}\left(\mu_{40}-\mu_{20}^2\right)$ and

$$\mathrm{MSE}(d_1)\mathrm{min} = \mathrm{MSE}(d_{1g})\mathrm{min}$$

$$= \frac{1}{n}\left(\mu_{40}-\mu_{20}^2\right)-\frac{\left\{\frac{2\mu_{21}\mu_{03}}{n\mu_{02}}\left(\mu_{20}-\frac{\mu_{22}}{\mu_{02}}\right)-\frac{(\beta_2-1)\mu_{21}^2}{n}-\frac{\mu_{02}}{n}\left(\mu_{20}-\frac{\mu_{22}}{\mu_{02}}\right)^2\right\}}{\left\{(\beta_2-1)\mu_{02}-\frac{\mu_{03}^2}{\mu_{02}^2}\right\}} \tag{4.1}$$

And

$$\mathrm{MSE}(d_2)_{\min} = \mathrm{MSE}(d_{2g})_{\min}$$

$$= \frac{1}{n}\left(\mu_{40}-\mu_{20}^2\right)-\frac{\left\{\left(\frac{2\mu_{21}\mu_{03}}{n\mu_{02}}\right)\left(\mu_{20}-\frac{\mu_{22}}{\mu_{02}}\right)+\frac{\mu_{02}}{n}\left(\mu_{20}-\frac{\mu_{22}}{\mu_{02}}\right)^2+\frac{(\beta_2-1)}{n}\mu_{21}^2\right\}}{\left\{\mu_{02}(\beta_2-1)-\left(\frac{\mu_{03}}{\mu_{02}}\right)^2\right\}} \tag{4.2}$$

Comparative studies regarding their efficiency over the usual conventional unbiased estimator are carried out with the help of a numerical illustration.

Considering the data given in Cochran (1977) dealing with Paralytic Polio Cases 'Placebo' (Y) group, Paralytic Polio Cases in not inoculated group (X), computations of required values of μ_{rs} have been done and comparisons are made for a simple random sample of size n. For the data considered, we have

$$n = 34,\ \overline{Y} = 2.58,\ \overline{X} = 8370.6$$
$$\mu_{20} = 9.8894,\ \mu_{02} = 7.1865882 \times 10^{7}$$
$$\mu_{30} = 47.015235,\ \mu_{03} = 1.4510955 \times 10^{12}$$
$$\mu_{40} = 421.96088,\ \mu_{04} = 4.5961952 \times 10^{16}$$

$$\mu_{21} = 93.464705 \times 10^{3}\ \mu_{12} = 3.443287 \times 10^{8}$$
$$\mu_{11} = 19.435294 \times 10^{3}\ \mu_{22} = 3.0156658 \times 10^{9}$$

Using above values, we have

Mean Square Error of usual conventional unbiased estimator = 9.534136695 and

$$\text{MSE}(d_1) = \text{MSE}(d_{1g}) = \text{MSE}(d_2)$$
$$= \text{MSE}(d_{2g}) = 5.512540843,$$

Hence the percent relative efficiency (PRE) of the proposed estimators d_1, d_{1g}, d_2, and d_{2g} over the usual conventional estimator are given by

$$\text{PRE}(d_1) = \text{PRE}(d_{1g}) = \text{PRE}(d_2)$$
$$= \text{PRE}(d_{2g}) = 172.9535792,$$

showing that the proposed estimators d_1, d_{1g}, d_2, and d_{2g} are more efficient with highly significant percent relative efficiency over the usual conventional unbiased estimator of the population variance.

5. CONCLUSIONS

We have derived new sampling estimators of population variance using auxiliary information in the form of mean and variance both, the bias and mean square error equations are obtained. Using these equations, MSE of proposed estimators are compared with the traditional estimator in theory and shown that the proposed estimators have smaller MSE than the traditional one.

ACKNOWLEDGEMENTS

The authors are thankful to the referees and the Editor- in- chief for providing valuable suggestions regarding improvement of the paper.

REFERENCES

1. Cochran, W. C. (1977) – 'Sampling Techniques', 3rd Edition, John Wiley and Sons, New York.
2. Das, A. K. and Tripathi, T. P. (1978) – Use of auxiliary information in estimating the finite population variance. Sankhya, Vol. 40, Series C, 139-148.
3. Singh, R. K., Zaidi, S. M. H. and Rizvi, S. A. H (1995) – Estimation of finite population variance using auxiliary information, Journ. of Statistical Studies, Vol. 15, 17-28.
4. Singh, R. K., Zaidi, S. M. H. and Rizvi, S. A. H.(1996) – Some estimators of finite population variance using information on two auxiliary variables., Microelectron, Reliab., Vol. 36, No. 5, 667-670.
5. Srivastava, S. K. and Jha, J. J. H. S. (1980) – A class of estimators using auxiliary information for estimating finite population variance, Sankhya, Vol. 40, Series C, 87-96.

Chapter 8

Statistical Methods of SNP Data Analysis and Applications

Alexander Balatskiy[2], Larisa Samokhodskaya[2], Vsevolod Tkachuk[2]

[1]*Faculty of Mathematics and Mechanics, Moscow State University, Moscow, Russia*

[2]*Faculty of Basic Medicine, Moscow State University, Moscow, Russia*

ABSTRACT

We develop various statistical methods important for multidimensional genetic data analysis. Theorems justifying application of these methods are established. We concentrate on the multifactor dimensionality reduction, logic regression, random forests, stochastic gradient boosting along with their new modifications. We use complementary approaches to study the risk of complex diseases such as cardiovascular ones. The roles of certain combinations of single nucleotide polymorphisms and non-genetic risk factors are examined. To perform the data analysis concerning the coronary heart disease and myocardial infarction the Lomonosov Moscow State University supercomputer "Chebyshev" was employed.

Keywords: Genetic Data Statistical Analysis; Multifactor Dimensionality Reduction; Ternary Logic Regression; Random Forests; Stochastic Gradient Boosting; Independent Rule; Single Nucleotide Polymorphisms; Coronary Heart Disease; Myocardial Infarction

1. INTRODUCTION

In the last decade new high-dimensional statistical methods were developed for the data analysis (see, e.g., [1]). Special attention was paid to the study of genetic models (see, e.g., [2-4]). The detection of genetic susceptibility to complex diseases (such as diabetes and others) has recently drawn much attention in leading research centers. It is well-known that such diseases can be provoked by variations in different parts of the DNA code which are responsible for the formation of certain types of proteins. One of the most common individual's DNA variations is a single nucleotide polymorphism (SNP), i.e. a nucleotide change in a certain fragment of genetic code (for some percentage of population). Quite a number of recent studies (see, e.g., [5,6] and references therein) support the paradigm that certain combinations of SNP can increase the complex disease risk whereas separate changes may have no dangerous effect.

There are two closely connected research directions in genomic statistics. The first one is aimed at the disease risk estimation assuming the genetic portrait of a person is known (in turn this problem involves estimation of disease probability and classification of genetic data into high and low risk domains). The second trend is to identify relevant combinations of SNPs having the most significant influence, either pathogenic or protective.

In this paper we propose several new versions of statistical methods to analyze multidimensional genetic data, following the above-mentioned research directions. The methods developed generalize the multifactor dimensionality reduction (MDR) and logic regression (LR). We employ also some popular machine learning methods (see, e.g., [2]) such as random forests (RF) and stochastic gradient boosting (SGB).

Ritchie et al. [7] introduced MDR as a new method of analyzing gene-gene and gene-environment interactions. Rather soon the method became very popular. According to [8], since the first publication more than 200 papers applying MDR in genetic studies were written.

LR was proposed by Ruczinski et al. in [9]. Further generalizations are given in [6,10] and other works. LR is based on the classical binary logistic regression and the exhaustive search for relevant predictor combinations. For genetic analysis it is convenient to use explanatory variables taking 3 values. Thus we employ ternary variables and ternary logic regression (TLR), whereas the authors of the above-mentioned papers employ binary ones.

RF and SGB were initiated by Breiman [11] and Friedman [12] respectively. They belong to ensemble methods which combine multiple predictions from a certain base algorithm to obtain better predictive power. RF and SGB were successfully applied to genetics data in a number of papers (see [2,13] and references therein).

We compare various approaches on the real datasets concerning coronary heart disease (CHD) and myocardial infarction (MI). Each approach (MDR, TLR and machine learning) is characterized by its own way of constructing disease prediction algorithms. For each method one or several prediction algorithms admitting the least estimated prediction error are found (a typical situation is that there are several ones with almost the same estimated prediction error). These prediction algorithms provide a way to determine the domains where the disease risk is high or low. It is also possible to select combinations of SNPs and non-genetic risk factors influencing the liability to disease essentially. Some methods allow to present such combinations immediately. Other ones, which employ more complicated forms of dependence between explanatory and response variables, need further analysis based on modifications of permutation tests. New software implementing the mentioned statistical methods has been designed and used.

This work was started in 2010 in the framework of the general MSU project headed by Professors V. A. Sadovnichy and V. A. Tkachuk (see [14]). MSU supercomputer "Chebyshev" was employed to perform data analysis.

The rest of the paper is organized as follows. In Section 2 we discuss various statistical methods and prove theorems justifying their applications. Section 3 is devoted to analysis of CHD and MI datasets. Section 4 contains conclusions and final remarks.

2. METHODS

We start with some notation. Let N be the number of patients in the sample and let the vector $X^j = (X_1^j, \cdots, X_n^j)$ consist of genetic (SNP) and nongenetic risk factors of individual j $(j = 1, \cdots, N)$. Here n is the total number of factors and X_i^j is the value of the i-th variable (genetic or non-genetic factor) of individual j. These variables are called explanatory variables or predictors. If X_i^j stands for a genetic factor (characterizes the i-th SNP of individual j) we set

$$X_i^j = \begin{cases} 0, & \text{SNP is homozygous for dominant allele,} \\ 1, & \text{SNP is heterozygous} \\ 2, & \text{SNP is homozygous for recessive allele.} \end{cases}$$

For biological background we refer, e.g., to [15].

We assume that non-genetic risk factors also take no more than three values, denoted by 0, 1 and 2. For example, we can specify a presence or absence of obesity (or hypercholesterolemia etc.) by the values 1 and 0 respectively. If a non-genetic factor takes more values (e.g., blood pressure), we can divide individuals into three groups according to its values.

Further on $X_1^j,\cdots,X_m^j$ stand for genetic data and $X_{m+1}^j,\cdots,X_n^j$ for non-genetic risk factors. Let a binary variable Y^j (response variable) be equal to 1 for a case, i.e. whenever individual j is diseased, and to –1 otherwise (for a control). Set

$$\xi=\left(\xi^1,\cdots,\xi^N\right)$$

where

$$\xi^j=(X^j,Y^j),\ \ j=1,\cdots,N\ .$$

Suppose $\xi^1,\cdots,\xi^N$ are i.i.d. random vectors. Introduce a random vector (X,Y) independent of ξ and having the same law as ξ^1. All random vectors (and random variables) are considered on a probability space $(\Omega,\mathcal{F},\mathrm{P})$, E denotes the integration w.r.t. P.

The main problem is to find a function in genetic and non-genetic risk factors describing the phenotype (that is the individual being healthy or sick) in the best way.

2.1. Prediction Algorithms

Let $\mathbf{X}:=\{0,1,2\}^n$ denote the space of all possible values of explanatory variables. Any function $f:\mathbf{X}\to\{-1,1\}$ is called a theoretical prediction function. Define the balanced or normalized prediction error for a theoretical prediction function f as

$$Err(f):=\mathrm{E}\left|Y-f(X)\right|\psi(Y)$$

where the penalty function $\psi:\{-1,1\}\to\mathbf{R}_+$. Obviously

$$\begin{aligned}Err(f)=&\,2\psi(-1)\mathrm{P}\left(f(X)=1,Y=-1\right)\\&+2\psi(1)\mathrm{P}\left(f(X)=-1,Y=1\right).\end{aligned} \tag{1}$$

Clearly $Err(f)$ depends also on the law of (X,Y) but we simplify the notation. Following [8,16] we put

$$\psi(y) = \frac{1}{4\mathrm{P}(Y=y)}, \quad y \in \{-1,1\},$$

where the trivial cases $\mathrm{P}(Y=-1)=0$ and $\mathrm{P}(Y=1)=0$ are excluded. Then

$$Err(f) = \frac{1}{2}\mathrm{P}\left(f(X)=1|Y=-1\right) + \frac{1}{2}\mathrm{P}(f(X)=-1|Y=1). \tag{2}$$

If $\mathrm{P}(Y=-1)=\mathrm{P}(Y=1)=1/2$ a sample is called balanced and one has $Err(f)=\mathrm{E}|Y-f(X)|/2$. Therefore in this case $Err(f)$ equals the classification error $\mathrm{P}(Y \neq f(X))$. In general,

$$Err(f) = \frac{1}{2}\mathrm{E}\left|Y^* - f(X^*)\right|.$$

with (X^*,Y^*) having the distribution

$$\mathrm{P}\left(X^* = x, Y^* = y\right) = \frac{1}{2}\mathrm{P}\left(X = x|Y = y\right),$$

$$(x,y) \in X \times \{-1,1\}$$

The reason to consider this weighted scheme is that a misclassification in a more rare class should be taken into account with a greater weight. Otherwise, if the probability of disease $\mathrm{P}(Y=1)$ is small, then the trivial function $f(x) \equiv -1$ may have the least prediction error.

It is easy to prove that the optimal theoretical predicttion function minimizing the balanced prediction error is given by

$$f^*(x) = \begin{cases} 1, & p(x) > \mathrm{P}(Y=1), \\ -1, & \text{otherwise}, \end{cases} \tag{3}$$

where

$$p(x) = \mathrm{P}\left(Y = 1|X = x\right), \quad x \in X. \tag{4}$$

Then each multilocus genotype (with added non-genetic risk factors) $x \in X$ is classified as high-risk if $f^*(x)=1$ or low-risk if $f^*(x)=-1$.

Since $p(x)$ and $\mathrm{P}(Y=1)$ are unknown, the immediate application of (3) is not possible. Thus we try to find an approximation of unknown function f^* using a prediction algorithm that is a function

$$f_{PA} = f_{PA}\left(x, \xi(S)\right)$$

with values in $\{-1,1\}$ which depends on $x \in X$ and the sample

$$\xi(S) = \{\xi^j, j \in S\}$$

where

$$S \subset \{1, \cdots, N\}. \tag{5}$$

The simplest way is to employ formula (3) with $p(x)$ and $\mathrm{P}(Y=1)$ replaced by their statistical estimates. Consider

$$\hat{p}\left(x, \xi(S)\right) = \frac{\sum_{j \in S} I\{Y^j = 1, X^j = x\}}{\sum_{j \in S} I\{X^j = x\}}, \; x \in X, \tag{6}$$

and take

$$\hat{\mathrm{P}}_S\left(Y = 1\right) = \frac{1}{\# S} \sum_{j \in S} I\{Y^j = 1\} \tag{7}$$

where $I\{A\}$ stands for the indicator of an event A and #D denotes the cardinality of a finite set D.

Along with (7) we consider

$$\hat{\mathrm{P}}_S\left(Y = 1 \middle| X \in C\right) = \frac{\sum_{j \in S} I\{Y^j = 1, X^j \in C\}}{\sum_{j \in S} I\{X^j \in C\}}, \tag{8}$$

for $C \subset X$. Thus (6) is a special case of (8) for $C = \{x\}$ with $x \in X$. Note that a more difficult way is to search for the estimators of f^* using several subsamples of ξ.

For a given prediction algorithm f_{PA}, keeping in mind (2), we set

$$Err\left(f_{PA}\left(\cdot,\xi(S)\right)\right) = \frac{1}{2}\sum_{y\in\{-1,1\}} \mathrm{P}\left(f_{PA}\left(X,\xi(S)\right)\neq y \middle| Y=y\right). \tag{9}$$

If one deals with too many parameters, overfitting is likely to happen, i.e. estimated parameters depend too much on the given sample. As a result the constructed estimates give poor prediction on new data. On the other hand, application of too simple model may not capture the studied dependence structure of various factors efficiently. However the trade-off between the model's complexity and its predictive power allows to perform reliable statistical inference via new model validation techniques (see, e.g., [17]). The main tool of model selection is the cross-validation, see, e.g., [16]. Its idea is to estimate parameters by involving only a part of the sample (training sample) and afterwards use the remaining observations (test sample) to test the predictive power of the obtained estimates. Then an average over several realizations of randomly chosen training and test samples is taken, see [18].

As the law of (X,Y) is unknown, one can only construct an estimate $\widehat{Err}\left(f_{PA}\left(\cdot,\xi(S)\right)\right)$ of $Err\left(f_{PA}\left(\cdot,\xi(S)\right)\right)$. In Section 3 we use the estimated prediction error (EPE) of a prediction algorithm f_{PA} which is based on K-fold cross-validation $(K>1)$ and has the form

$$\widehat{Err}_K\left(f_{PA}(\cdot,\xi),\xi\right) = \frac{1}{2}\sum_{y\in\{-1,1\}}\frac{1}{K}\sum_{k=1}^{K}\frac{\sum_{(k)} I\left\{f_{PA}\left(X^j,\xi\left(\overline{S_k}\right)\right)\neq y, Y^j=y\right\}}{\sum_{(k)} I\left\{Y^j=y\right\}} \tag{10}$$

where the sum $\sum_{(k)}\cdot$ is taken over j belonging to

$$S_k=\left\{(k-1)[N/K]+1,\cdots, k[N/K]I\{k<K\}+NI\{k=K\}\right\}, \tag{11}$$

$\overline{S_k}=\{1,\cdots,N\}\setminus S_k$ and $[a]$ is the integer part of $a\in\mathbf{R}$.

Let $\{k_1,\cdots,k_r\}\subset\{1,\cdots,n\}$. Introduce

$$C_{k_1,\cdots,k_r}(x)=\left\{u=(u_1,\cdots,u_n)\in\mathbb{X}:u_{k_i}=x_{k_i},\ i=1,\cdots,r\right\}.$$

The next result provides wide sufficient conditions for consistency of estimate (10).

Theorem 1. Suppose there exist a subset $U \subset X$ and a subset $\{k_1,\cdots,k_r\} \subset \{1,\cdots,n\}$ such that the following holds:

1) For each $x \in X$ and any finite dimensional vector v with components belonging to $X \times \{-1,\ 1\}$, functions $f_{PA}(\cdot,v)$ and f are constant on $C_{k_1,\cdots,k_r}(x)$.

2) For each $x \in U$ and any $W_N \subset \{1,\cdots,N\}$ with $\#W_N \to \infty$, one has $f_{PA}(x,\xi(W_N)) \to f(x)$ a.s. when $N \to \infty$.

3) $\mathrm{P}(Y=1|X_{k_1}=x_{k_1},\ldots,X_{k_r}=x_{k_r})=\mathrm{P}(Y=1)$ if $x \in X \setminus U$.

4) f is constant on $X \setminus U$.

Then $\hat{Err}_K(f_{PA}(\cdot,\xi),\xi) \to Err(f)$ a.s., $N \to \infty$.

Remark. If we replace condition 3 of Theorem 1 by a more restrictive assumption 3') $\mathrm{P}(Y=1|X=x)=\mathrm{P}(Y=1)$ for all $x \in X \setminus U$. then we can take $\{k_1,\cdots,k_r\}=\{1,\cdots,n\}$ to remove condition 1.

Proof. The proof is based on the following Lemma. Let $\{(Z_j^{(m)},Y_j^{(m)}),\ 1 \le j \le m, m \in N\}$ be an array of rowwise independent random elements distributed as (Z,Y), where Z takes values in a finite set Z and Y takes values in $\{-1,1\}$. Assume that $\{f_m(z),\ m \in N,\ z \in Z\}$ is an array of random variables with values in $\{-1,1\}$.

Suppose there exists $U \subset Z$ such that the following conditions hold:

1) $f_m(z) \to f(z)$ a.s. for all $z \in U$, as $m \to \infty$, where a nonrandom function $f: Z \to \{-1,1\}$.

2) $\mathrm{P}(Y=1|Z=z)=\mathrm{P}(Y=1)$ if $z \in Z \setminus U$.

3) f is constant on $Z \backslash U$.

Then, as $m \to \infty$,

$$\frac{1}{2} \sum_{y \in \{-1,1\}} \frac{\sum_{j=1}^{m} I\left\{f_m\left(Z_j^{(m)}\right) \neq y, Y_j^{(m)} = y\right\}}{\sum_{j=1}^{m} I\left\{Y_j^{(m)} = y\right\}} \to Err(f) \text{ a.s.} \tag{12}$$

Proof of Lemma. Set $Q_m(y) = \sum_{j=1}^{m} I\left\{Y_j^{(m)} = y\right\}$ and define events

$$A_j^{(m)}(y) = \left\{f_m\left(Z_j^{(m)}\right) \neq y, Y_j^{(m)} = y\right\},$$

$$B_j^{(m)}(y) = \left\{f(Z_j^{(m)}) \neq y, Y_j^{(m)} = y\right\}.$$

Then the l.h.s. of (12) equals

$$\sum_{y \in \{-1,1\}} \frac{1}{2Q_m(y)} \sum_{j=1}^{m} I\left\{A_j^{(m)}(y)\right\}.$$

For $y \in \{-1,1\}$, we have

$$\begin{aligned} &\frac{1}{Q_m(y)} \sum_{j=1}^{m} I\left\{A_j^{(m)}(y)\right\} \\ &= \frac{1}{m\mathrm{P}\{Y = y\}} \sum_{j=1}^{m} I\left\{A_j^{(m)}(y)\right\} \\ &+ \frac{1}{m} \sum_{j=1}^{m} I\left\{A_j^{(m)}(y)\right\}\left(\frac{m}{Q_m(y)} - \frac{1}{\mathrm{P}(Y = y)}\right). \end{aligned} \tag{13}$$

The absolute value of the second term in the r.h.s. of (13) does not exceed $\left|m/(Q_m(y)) - 1/\mathrm{P}(Y = y)\right|$. and tends to 0 a.s. if $m \to \infty$. This statement follows from the strong law of large numbers for arrays (SLLNA), see [19].

Note that

$$\begin{aligned}
&\frac{1}{m}\sum_{j=1}^{m} I\left\{A_j^{(m)}(y)\right\} \\
&= \frac{1}{m}\sum_{j=1}^{m} I\left\{B_j^{(m)}(y)\right\} \\
&+ \frac{1}{m}\sum_{j=1}^{m} I\left\{Z_j^{(m)} \in U\right\}\left(I\left\{A_j^{(m)}(y)\right\} - I\left\{B_j^{(m)}(y)\right\}\right) \\
&+ \frac{1}{m}\sum_{j=1}^{m} I\left\{Z_j^{(m)} \notin U\right\}\left(I\left\{A_j^{(m)}(y)\right\} - I\left\{B_j^{(m)}(y)\right\}\right).
\end{aligned} \tag{14}$$

According to SLLNA the first term in the r.h.s. of (14) goes to $P(f(Z) \neq y, Y = y)$ a.s. We claim that the second term tends to 0 a.s. Indeed, the set Z is finite and the functions f_m f, take only two values. Therefore, by condition 1, for almost all $\omega \in \Omega$, there exists $N_1 = N_1(\omega)$ such that $f_m(z) = f(z)$ for all $z \in U$ and $m > N_1$. Hence second term in the r.h.s. of (14) equals 0 for all $m > N_1$, which proves the claim. Thus, it remains to estimate the third term.

In view of condition 3, w.l.g. we may assume that $f(z) = -1$ for $z \in Z \setminus U$. Then we obtain

$$\begin{aligned}
V_m &:= \sum_{y \in \{-1,1\}} \sum_{j=1}^{m} \frac{I\left\{Z_j^{(m)} \notin U\right\}\left(I\left\{A_j^{(m)}(y)\right\} - I\left\{B_j^{(m)}(y)\right\}\right)}{m\mathrm{P}(Y = y)} \\
&= \frac{1}{m}\sum_{j=1}^{m} I\left\{Z_j^{(m)} \notin U\right\} I\left\{f_m\left(Z_j^{(m)}\right) = 1\right\} R_j^{(m)} \\
&= \frac{1}{m}\sum_{z \in Z \setminus U} I\left\{f_m(z) = 1\right\} \sum_{j=1}^{m} I\left\{Z_j^{(m)} = z\right\} R_j^{(m)}
\end{aligned} \tag{15}$$

where

$$R_j^{(m)} = \frac{I\left\{Y_j^{(m)} = -1\right\}}{\mathrm{P}(Y = -1)} - \frac{I\left\{Y_j^{(m)} = 1\right\}}{\mathrm{P}(Y = 1)}.$$

SLLNA and condition 2 imply that, for $z \in Z \setminus U$ and $y \in \{-1, 1\}$, we have

$$\sum_{j=1}^{m} \frac{I\{Z_j^{(m)} = z\} I\{Y_j^{(m)} = y\}}{m\mathrm{P}(Y = y)}$$
$$\to \frac{\mathrm{P}(Z = z, Y = y)}{\mathrm{P}(Y = y)} = \mathrm{P}(Z = z)$$

almost surely. Therefore, for almost all $\omega \in \Omega$, there exists $N_2 = N_2(\omega)$ such that

$$\left| \frac{1}{m} \sum_{j=1}^{m} I\{Z_j^{(m)} = z\} R_j^{(m)} \right| < \varepsilon$$

for all $z \in Z \setminus U$ and $m > N_2$. Using the last estimate and (15) we finally get that, for $m > N_2$,

$$|V_m| \le \sum_{z \in Z \setminus U} \varepsilon\, I\{f_m(z) = 1\} \le \varepsilon \cdot \#Z.$$

Hence $V_m \to 0$ a.s. if $m \to \infty$. Combining (12)-(16) we obtain the desired result.

Let us return to the proof of Theorem 1. Fix $1 \le k \le K$ and take

$$f_m(z) := f_{PA}\left(x, \xi\left(\overline{S_k}\right)\right)$$

where $z \in Z := \{0,1,2\}^r$, $m := \#S_k$, S_k is introduced in (11), and x is any element of X with $(x_{k_1}, \cdots, x_{k_r}) = z.$ By condition 1 of Theorem 1, f_m is well defined.

Applying Lemma to arrays

$$\left\{\left(Z_j^{(m)}, Y_j^{(m)}\right),\ 1 \le j \le m\right\} := \left(\left(X_{k_1}^j, \ldots, X_{k_r}^j\right), Y^j\right), j \in S_k\right\}$$

and $\{f_m(z),\ z \in Z\}$, we obtain that almost surely

$$\frac{1}{2} \sum_{y \in \{-1,1\}} \frac{\Sigma_{(k)} I\left\{f_{PA}\left(X^j, \xi\left(\overline{S_k}\right)\right) \ne y, Y^j = y\right\}}{\Sigma_{(k)} I\{Y^j = y\}} \to Err(f)$$

as $\#S_k \to \infty$. Thus

$$\hat{Err}_K\left(f_{PA}(\cdot,\xi),\xi\right)\to\frac{1}{K}\sum_{k=1}^{K}Err(f)=Err(f)$$

when $N\to\infty$, which completes the proof of Theorem 1.

An important problem is to make sure that the predicttion algorithm f_{PA} gives statistically reliable results. The quality of an algorithm is determined by its predicttion error (9) which is unknown and therefore the inference is based on consistent estimates of this error. Clearly the high quality of an algorithm means that it captures the dependence between predictors and response variables, so the error is made more rarely than it would be if these variables were independent. Consider a null hypothesis H_0 that X and Y are independent. If they are in fact dependent, then for any reasonable prediction algorithm f_{PA} an appropriate test procedure involving f_{PA} should reject H_0 at the significance level α , e.g., 5%. This shows that the results of the algorithm could not be obtained by chance. For such a procedure, we take a permutation test which allows to find the Monte Carlo estimate $\widehat{p}=\widehat{F}\left(\widehat{Err}_K\left(f_{PA},\xi\right)\right)$ (see [20]) of the true p-value $p=F\left(\hat{Err}_K\left(f_{PA},\xi\right)\right)$.

Here F is the cumulative distribution function (c.d.f.) of $\hat{Err}_K$ under H_0 and $\widehat{F}=\widehat{F}(z)$ is the corresponding empirical c.d.f. We reject H_0 if $\widehat{p}<\alpha.$ For details we refer to [21].

Now we pass to the description of various statistical methods and their applications (in Section 3) to the cardiovascular risk detection.

2.2. Multifactor Dimensionality Reduction

MDR is a flexible non-parametric method of analyzing gene-gene and gene-environment interactions. MDR does not depend on a particular inheritance model. We give a rigorous description of the method following ideas of [7] and [8].

To calculate the balanced error of a theoretical predicttion function we use formula (2). Note that the approach based on penalty functions is not the only possible. Nevertheless it outperforms substantially other approaches involving overand undersampling (see [8]).

As mentioned earlier, the probability p(x) introduced in (4) is unknown. To find its estimate one can apply maximum likelihood approach assuming that the random variable $I\{Y=1\}$ conditionally on $X=x$ has a Bernoulli distribution with unknown parameter $p(x)$. Then we come to (6).

A direct calculation of estimate (6) with exhaustive search over all possible values of x is highly inefficient, since the number of different values of x grows exponentially with number of risk factors. Moreover, such a search leads to overfitting. Instead, it is usually supposed that $p(x)$ depends non-trivially not on all, but on certain variables x_i. That is, there exist $l \in \mathbf{N},\ l<n,$ and a vector $(k_1^*,\cdots,k_l^*),$ where $1 \le k_1^* < \cdots < k_l^* \le n,$ such that for each $x=(x_1,\cdots,x_n)\in \mathcal{X},$ the following relation holds:

$$p(x)=\mathrm{P}\left(Y=1 \middle| X_{k_1^*}=x_{k_1^*},\cdots,X_{k_l^*}=x_{k_l^*}\right) \tag{17}$$

In other words only few factors influence the disease and other ones can be neglected. A combination of indices $(k_1^*,\cdots,k_l^*),$ in formula (17) having minimal l is called the most significant.

For $x\in\mathcal{X}$ and indices $\{k_1,\cdots,k_r\}\subset\{1,\cdots,n\},$ set

$$f_{k_1,\cdots,k_r}(x) = \begin{cases} 1, & \mathrm{P}(Y=1 \mid X_{k_1}=x_{k_1},\ldots,X_{k_r}=x_{k_r}) > \mathrm{P}(Y=1), \\ -1, & \text{otherwise.} \end{cases}$$

Consider the estimate

$$\hat{f}_{k_1,\cdots,k_r}(x,\xi(S)) := \begin{cases} 1, & \hat{\mathrm{P}}_S(Y=1 \mid X\in C_{k_1,\cdots,k_r}(x)) > \hat{\mathrm{P}}_S(Y=1), \\ -1, & \text{otherwise,} \end{cases} \tag{18}$$

where S is introduced in (5).

Theorem 2. Let $(k_1^*,\cdots,k_l^*)$ be the most significant combination. Then for any fixed $\{k_1,\cdots,k_r\}\subset\{1,\cdots,n\}$ one has 1) $Err\left(f_{k_1^*,\cdots,k_l^*}\right)\le Err\left(f_{k_1,\cdots,k_r}\right);$

2) $\widehat{Err}_K\left(\widehat{f}_{k_1,\cdots,k_r}\right)$ is a strongly consistent asymptotically unbiased estimate of $Err\left(f_{k_1,\cdots,k_r}\right)$ as $N \to \infty$;

3) for any $\varepsilon,\ \delta > 0$ and all N large enough

$$\mathrm{P}\left(\widehat{Err}_K\left(\widehat{f}_{k_1^*,\cdots,k_l^*}\right) < \widehat{Err}_K\left(\widehat{f}_{k_1,\cdots,k_r}\right) + \varepsilon\right) > 1 - \delta.$$

Proof. 1) It follows from (17) that $f_{k_1^*,\cdots,k_l^*}$ coinsides with function f^* (see (3)) which has the minimal balanced prediction error.

2) Let us verify the conditions of Theorem 1 for $f_{PA}\left(x, \xi(S)\right) := \widehat{f}_{k_1,\cdots,k_r}\left(x, \xi(S)\right)$. Condition 1 follows from the definition of $\widehat{f}_{k_1,\cdots,k_r}\left(x, \xi(S)\right)$. Further, put

$$U := \left\{x \in X : \mathrm{P}\left(Y = 1 \middle| C_{k_1,\ldots,k_r}(x)\right) \neq \mathrm{P}(Y = 1)\right\} \tag{19}$$

where $C_{k_1,\cdots,k_r}$ was introduced in Section 2.1 before Theorem 1. We claim that, for each $x \in U$ and any $W_N \subset \{1, \cdots, N\}$ such that $\# W_N \to \infty$, the following relation holds:

$$\widehat{f}_{k_1,\cdots,k_r}\left(x, \xi(W_N)\right) \to f_{k_1,\cdots,k_r}(x) \text{ a.s. if } N \to \infty.$$

Indeed, assume that, for some $\varepsilon > 0,$

$$\mathrm{P}\left(Y = 1 \middle| X_{k_1} = x_{k_1}, \cdots, X_{k_r} = x_{k_r}\right) - \mathrm{P}(Y = 1) > \varepsilon.$$

SLLNA implies that

$$\widehat{\mathrm{P}}_{W_N}\left(Y = 1 \middle| X \in C_{k_1,\ldots,k_r}(x)\right) - \widehat{\mathrm{P}}_{W_N}(Y = 1)$$

converges a.s. to

$$\mathrm{P}\left(Y = 1 \middle| X_{k_1} = x_{k_1}, \cdots, X_{k_r} = x_{k_r}\right) - \mathrm{P}(Y = 1),$$

$N \to \infty$. Then, for almost all $\omega \in \Omega$, there exists $N_0 = N_0(\omega)$ such that

$$\widehat{\mathrm{P}}_{W_N}\left(Y=1 \mid X \in C_{k_1,\cdots,k_r}(x)\right)-\widehat{\mathrm{P}}_{W_N}\left(Y=1\right)>\varepsilon/2$$

for all $N > N_0$. Therefore, for all $N > N_0$, we have $\widehat{f}_{k_1,\cdots,k_r}\left(x,\xi(W_N)\right)=1=f_{k_1,\cdots,k_r}(x)$, which proves the claim. Thus condition 2 of Theorem 1 is met.

Conditions 3 and 4 of Theorem 1 follow from (19) and the definitions of $\widehat{f}_{k_1,\cdots,k_r}\left(x,\xi(S)\right)$ and $f_{k_1,\cdots,k_r}(x)$.

Since all conditions of Theorem 1 are satisfied, we have $\widehat{Err}_K\left(\widehat{f}_{k_1,\cdots,k_r}\right)\to Err\left(f_{k_1,\cdots,k_r}\right)$ a.s. and in mean (due to the Lebesgue theorem as $\widehat{Err}_K\left(\widehat{f}_{k_1,\cdots,k_r}\right)$ is bounded by 1).

3) Follows from 1) and 2). □

In view of this result it is natural to pick one or a few combinations of factors with the smallest EPEs as an approximation for the most significant combination.

The last step in MDR is to determine statistical significance of the results. Here we test a null hypothesis of independence between predictors X and response variable Y. This can be done via the permutation test mentioned in Section 2.1.

MDR method with "independent rule". We propose multifactor dimensionality reduction with "independent rule" (MDRIR) method to improve the estimate of probability $p(x)$. This approach is motivated by [22] which deals with classification of large arrays of binary data. The principal difficulty with employment of formula (6) is that the number of observations in numerator and denominator of the formula might be small even for large N (see, e.g., [23]). This can lead to inaccurate estimates and finally to a wrong prediction algorithm. Moreover, for some samples the denominator of (6) might equal zero.

The Bayes formula implies that $p(x)$ equals

$$\frac{\mathrm{P}\left(X=x \middle| Y=1\right)\mathrm{P}\left(Y=1\right)}{\mathrm{P}\left(X=x \middle| Y=1\right)\mathrm{P}\left(Y=1\right)+\mathrm{P}\left(X=x \middle| Y=-1\right)\mathrm{P}\left(Y=-1\right)} \tag{20}$$

where the trivial cases $\mathrm{P}(Y=-1)=0$ and $\mathrm{P}(Y=1)=0$ are excluded. Substituting (20) into (3) we obtain the following expression for the prediction function:

$$f^*(x) = \begin{cases} 1, & \mathrm{P}(X = x \mid Y = 1) > \mathrm{P}(X = x \mid Y = -1), \\ -1, & \text{otherwise.} \end{cases} \tag{21}$$

As in standard MDR method described above, we assume that formula (17) holds. It was proved in [22] that for a broad class of models (e.g., Bahadur and logit models) the conditional probability

$$\mathrm{P}\left(X_{k_1} = x_1, \cdots, X_{k_r} = x_r \middle| Y = y\right)$$

where $y = \pm 1,$ can be estimated in the following way:

$$\hat{\mathrm{P}}_S\left(X_{k_1} = x_1, \ldots, X_{k_r} = x_r \middle| Y = y\right) := \prod_{i=1}^{r} \hat{\mathrm{P}}_S\left(X_{k_i} = x_i \middle| Y = y\right), \tag{22}$$

here (cf. (8))

$$\hat{\mathrm{P}}_S\left(X_{k_i} = x \middle| Y = y\right) = \frac{\sum_{j \in S} I\left\{X_{k_i}^j = x, Y^j = y\right\}}{\sum_{j \in S} I\left\{Y^j = y\right\}}. \tag{23}$$

Combining (17) and (21)-(23) we find the desired estimate of $f^*(x)$.

A number of observations in numerator and denominator of (23) increases considerably comparing with (18). It allows to estimate the conditional probability more precisely whenever the estimate introduced in (22) is reasonable. For instance, sufficient conditions justifying the application of (22) are provided in [22, Cor.5.1]. MDRIR might have some advantage over MDR in case when the size l of the most significant combination $\left(k_1^*, \cdots, k_l^*\right)$ is large. However, MDR for small l can demonstrate better behavior than MDRIR.

Thus, as opposed to standard MDR method, MDRIR uses alternative estimates of conditional probabilities. All other steps (prediction algorithm construction, EPE calculation) remain the same. As far as we know, this modification of MDR has not been applied before. It is based on a combination of the original MDR method (see [7]) and the ideas of [22].

2.3. Ternary Logic Regression

LR is a semiparametric method detecting the most significant combinations of predictors as well as estimating the conditional probability of the disease.

Let $p^*(x) = \mathrm{P}(Y^* = 1 | X^* = x)$ (where $x \in \mathcal{X}$) be the conditional probability of a disease defined in normalized sample, where (X^*, Y^*) was introduced in Section 2.1. Note that formula (3) can be written as follows:

$$f^*(x) = \begin{cases} 1, & p^*(x) > 1/2, \\ -1, & \text{otherwise.} \end{cases}$$

We suppose that trivial situations when $p^*(x) \in \{0, 1\}$ do not occur and omit them from the consideration. To estimate $p^*(x)$ we pass to the logistic transform

$$q^*(x) = \lambda(p^*(x)) \tag{24}$$

where $\lambda(z) = \log(z/(1-z)),\ z \in (0,1),$ is the inverse logistic function. The logistic function equals $\Lambda(t) = (1 + e^{-t})^{-1}$, $t \in \mathbf{R}.$ Note that we are going to estimate the unknown disease probability with the help of linear statistics with appropriately selected coefficients. Therefore it is natural to avoid restrictions on possible values of the function estimated. Thus the logistic transform is convenient, as $p^*(x) \in (0, 1)$ for $x \in \mathcal{X},$ while $q^*(x)$ can take all real values.

Consider a class $\mathcal{G}$ of all real-valued functions in ternary variables $x_1, \cdots, x_n$. We call a model of the dependence between the disease and explanatory variables any subclass $M \subset \mathcal{G}$. Set

$$\hat{\psi}(y, \xi(S)) = \frac{1}{4\hat{\mathrm{P}}_S(Y = y)},\ y \in \{-1, 1\},$$

with $\hat{\mathrm{P}}_S(Y = y)$ appearing in (7). Define the normalized smoothed score function

$$L(h, \xi(S)) = \frac{1}{\#S} \sum_{j \in S} \varphi(-Y^j h(X^j)) \hat{\psi}(Y^j, \xi(S)) \tag{25}$$

where S was introduced in (5), $\varphi(t) = \log_2(1 + e^t)$ for $t \in \mathbf{R},$ and $h \in M$. In contrast to previous works our version of LR (more precisely, TLR) scheme involves normalization (cf. (1)), i.e. taking the observations with weights dependent on the proportion of cases and controls in subsample $\xi(S)$. An easy computation yields that $\arg\min_{h\in M} L(h,\xi(S))$ equals $\arg\max_{h\in M}$ of the function

$$\frac{1}{\#S}\sum_{j\in S}\left(\frac{I\{Y^j=1\}}{2\hat{P}_S(Y=1)}\log\gamma_j+\frac{I\{Y^j=-1\}}{2\hat{P}_S(Y=-1)}\log(1-\gamma_j)\right)$$

with $\gamma_j=\Lambda(h(X^j))$. That is, minimizing the score function is equivalent to the normalized maximum likelihood estimation of q^*.

The next theorem guarantees strong consistency of this estimation method whenever the model is correctly specified, i.e. $q^*\in M$. To formulate this result introduce $h(\cdot,\xi(S)) := \arg\min_{q\in M} L(q,\xi(S))$.

Theorem 3. Let $h_0\equiv 0$ $q^*\in M$, belong to M and

$$\min_{(x,y)\in\mathcal{X}\times\{-1,1\}} P(X=x|Y=y)>0.$$

Consider $W_N\subset\{1,\cdots,N\}$ and set

$$h_N(\cdot)=h(\cdot,\xi(W_N)).$$

Then $h_N(x)\to q^*(x)$ a.s. for all $x\in\mathcal{X}$ when $\#W_N\to\infty$. Moreover,

$Err_K(f_{PA}(\cdot,\xi),\xi)\to Err(f^*)$ a.s., $N\to\infty$

where $f_{PA}(\cdot,\xi)=2I\{\Lambda(h(\cdot,\xi))>1/2\}-1$.

Proof. At first we show that

$$h_N(x)<\varphi^{-1}\left(\frac{4}{P(X=x|Y=-1)}\right) \text{ a.s.}$$

for any $x\in\mathcal{X}$ and all $N>N_1=N_1(\omega)$. Put $l_N=\#W_N$. By definition

$$\frac{\varphi(h_N(x))}{4l_N}\sum_{j\in W_N}\frac{I\{X^j=x,Y^j=-1\}}{\hat{\mathrm{P}}_{W_N}(Y=-1)}\le L(h_N,\xi(W_N))$$
$$\le L(0,\xi(W_N))=\frac{1}{4l_N}\sum_{(j,y)\in W_N\times\{-1,1\}}\frac{I\{Y^j=y\}}{\hat{\mathrm{P}}_{W_N}(Y=y)}.$$

Using SLLNA, we get that a.s.

$$\max_{x\in\mathcal{X}}\left|\frac{1}{l_N}\sum_{j\in W_N}\frac{I\{X^j=x,Y^j=-1\}}{\hat{\mathrm{P}}_{W_N}(Y=-1)}\right.$$
$$\left.-\mathrm{P}(X=x|Y=-1)\right|\to 0.$$

Obviously

$$\frac{1}{l_N}\sum_{(j,y)\in W_N\times\{-1,1\}}\frac{I\{Y^j=y\}}{\hat{\mathrm{P}}_{W_N}(Y=y)}=2.$$

These relations imply the desired estimate of $h_N(x)$. Similarly we prove that

$$h_N(x)>-\varphi^{-1}\left(\frac{4}{\mathrm{P}(X=x|Y=1)}\right)$$

for any $x\in\mathcal{X}$ and all $N>N_2=N_2(\omega)$. Consequently, we see that $h_N\in M_C:=M\cap\{h:\|h\|_\infty\le C\}$ for $N>\max(N_1,N_2)$, here $\|h\|_\infty=\max_{x\in\mathcal{X}}|h(x)|$ and

$$C=\max_{(x,y)\in\mathcal{X}\times\{-1,1\}}\varphi^{-1}\left(\frac{4}{\mathrm{P}(X=x|Y=y)}\right).$$

If $h\in M_C$ then $|L(h,\xi(W_N))-\mathrm{E}\varphi(-Yh(X))\psi(Y)|$ is less than

$$\frac{\varphi(\|h\|_\infty)}{2}\sum_{y\in\{-1,1\}}\left(\left|\frac{1}{\hat{\mathrm{P}}_{W_N}(Y=y)}-\frac{1}{\mathrm{P}(Y=y)}\right|\right.$$
$$\left.+\sum_{x\in\mathcal{X}}\left|\sum_{j\in W_N}\frac{I\{X^j=x,Y^j=y\}}{l_N\mathrm{P}(Y=y)}-\frac{\mathrm{P}(X^j=x,Y^j=y)}{\mathrm{P}(Y=y)}\right|\right).$$

By SLLNA, we get that

$$L\left(h,\xi\left(W_N\right)\right)-\mathrm{E}\varphi\left(-Yh\left(X\right)\right)\psi\left(Y\right)\rightarrow 0 \qquad \text{a.s.} \quad (26)$$

uniformly over $\left\{h:\|h\|_\infty \le C\right\}$. Note also that

$$2\mathrm{E}\varphi\left(-Yh\left(X\right)\right)\psi\left(Y\right)=\mathrm{E}\varphi\left(-Y^*h\left(X^*\right)\right)$$
$$=-\mathrm{E}\log_2\left(\Lambda\left(h\left(X^*\right)\right)^{I\{Y^*=1\}}\left(1-\Lambda\left(h\left(X^*\right)\right)\right)^{1-I\{Y^*=1\}}\right).$$

By the (conditional) information inequality,

$$\mathrm{E}\log_2\left(\Lambda\left(h\left(X^*\right)\right)^{I\{Y^*=1\}}\left(1-\Lambda\left(h\left(X^*\right)\right)\right)^{1-I\{Y^*=1\}}\right).$$

attains its maximum over all functions h only at q^* introduced in (24). Under conditions of the theorem, $q^*\in M$. Therefore, by definitions of h_N and q^* we have

$$L\left(h_N,\xi\left(W_N\right)\right)\le L\left(q^*,\xi\left(W_N\right)\right),$$

$$\mathrm{E}\varphi\left(-Y^*h\left(X^*\right)\right)|_{h=h_N}\ge \mathrm{E}\varphi\left(-Y^*q^*\left(X^*\right)\right).$$

By (26) and SLLNA

$$L\left(h_N,\xi\left(W_N\right)\right)-\frac{1}{2}\mathrm{E}\varphi\left(-Y^*h\left(X^*\right)\right)|_{h=h_N}\rightarrow 0 \ \text{a.s.},$$

$$L\left(q^*,\xi\left(W_N\right)\right)\rightarrow\frac{1}{2}\mathrm{E}\varphi\left(-Y^*q^*(X^*)\right) \ \text{a.s.}$$

Hence

$$\mathrm{E}\varphi\left(-Y^*h\left(X^*\right)\right)|_{h=h_N}\rightarrow \mathrm{E}\varphi\left(-Y^*q^*\left(X^*\right)\right) \ \text{a.s.}$$

This is possible only when $h_N\left(x\right)\rightarrow q^*\left(x\right)$ a.s. for all $x\in\mathcal{X}$. Indeed, for almost all $\omega\in\Omega$, we can always take a subsequence $h_{N_k}=h_{N_k(\omega)}\left(\cdot,\omega\right)$ converging to some function $\mu=\mu\left(\cdot,\omega\right)$ with

$$\mathrm{E}\varphi\left(-Y^*\mu\left(X^*\right)\right)=\mathrm{E}\varphi\left(-Y^*q^*\left(X^*\right)\right).$$

Hence, by the information inequality, $\mu\left(\cdot,\omega\right)=q\left(\cdot\right)$.

To establish the second part of Theorem 3 we note that $f_{PA}(x,\xi)=2I\{\Lambda(h(x,\xi))>1/2\}-1$ converges a.s. to $f^*(x)=2I\{\Lambda(q^*(x))>1/2\}-1$ for all $x\in U$ where

$$U:=\{x\in\mathcal{X}:p^*(x)\neq 1/2\}$$
$$=\{x\in\mathcal{X}:\mathrm{P}(Y=1|X=x)\neq\mathrm{P}(Y=1)\}.$$

Then the conclusion follows from Remark after Theorem 1. The proof is complete.

A wide and easy to handle class of models is obtained by taking functions linear in variables $x_1,\cdots,x_n.$ or/and in their products. In turn these functions admit a convenient representation by elementary polynomials (EP). Recall that EP is a function T in ternary variables $x_1,\cdots,x_n.$ belonging to $\{0,1,2\}$ which can be represented as a finite sum of products $x_1^{u_1}\cdots x_n^{u_n}$ where $u_1,\cdots,u_n\in\mathbf{Z}_+$. The addition and multiplication of ternary variables is considered by modulo 3. Any EP can be represented as a binary tree in which knots (vertices which are not leaves) contain either the addition or multiplication sign, and each leaf corresponds to a variable. Different trees may correspond to the same EP, thus this relation is not one-to-one. However, it does not influence our problem, so we keep the notation T for a tree. A finite set of trees $F=(T_1,\cdots,T_s)$ is called a forest. For a tree T its complexity C(T) is the number of leaves. The complexity C(F) of a forest F is the maximal complexity of trees constituting F. It is clear that if $g\in\mathcal{G}$ then there exists $s\geq 1$ such that g has the form

$$g(x_1,\cdots,x_n)=\beta_0+\sum_{i=1}^{s}\beta_iT_i(x_1,\cdots,x_n), \tag{27}$$

here $\beta_0,\beta_1,\cdots,\beta_s\in\mathbf{R}$ and $T_1,\cdots,T_s$ are EP.

Let us say that function g belongs to a class $\mathcal{G}_r(s)$, where $s,r\in\mathbf{N}$, if there exist a decomposition (27) of g such that all trees T_i ($i=1,\cdots,s$) have complexity less or equal to r. We identify a function $g\in\mathcal{G}_r(s)$ with pair (F,β) where F is the corresponding forest and $\beta=(\beta_0,\cdots,\beta_s)$ is the vector of coefficients in (27).

Minimization of $L(h,\xi(S))$ defined by (25) over all functions $h\in M\subset G_r(s)$ is done in two alternating steps. First, we find the optimal value of b while F is fixed (which is the minimization of a smooth function in several variables) and then we search for the best F. The main difficulty is to organize this search efficiently. Here one uses stochastic algorithms, since the number of such forests increases rapidly when the complexity r grows. For $s\in\mathbf{N}$, a forest $F=(T_1,\cdots,T_s)$ and a subsample $\xi(S)$ (see (5)), consider a prediction algorithm f_{LR}^F setting

$$f_{\mathrm{LR}}^F(x)=\begin{cases}1, & \hat{h}(x)>0,\\ -1, & \text{otherwise,}\end{cases}$$

where $\hat{h}=(F,\hat{\beta})$ and

$$\hat{\beta}=\arg\min_{\beta} L\left(\beta_0+\sum_{j=1}^{s}\beta_j T_j(\cdot),\xi(S)\right).$$

Define also the normalized prediction error of a forest $F=(T_1,\cdots,T_s)$ as $\phi(F)=\widehat{Err}_K(f_{\mathrm{LR}}^F(\cdot,\xi),\xi)$.

A subgraph B of a tree T is called a branch if it is itself a binary tree (i.e. it can be obtained by selecting one vertex of T together with its offspring). The addition and multiplication signs standing in a knot of a tree are called operations, thus * stands for sum or product. Following [9], call the tree T' a neighbor of T if it is obtained from T via one and only one of the following transformations.

1) Changing one variable to another in a leaf of the tree T (variable change).
2) Replacing an operation in a knot of a tree T with another one, i.e. sum to product or vice versa (operator change).
3) Changing a branch of two leaves to one of these leaves (deleting a leaf).
4) Changing a leaf to a branch of two leaves, one of which contains the same variable as in initial leaf (splitting a leaf).
5) Replacing a branch $B_1 * B_2$ with the branch B_1 (branch pruning).

6) Changing a branch B to a branch $x_i * B$ (branch growing), here x_i is a variable.

We say that forests F and F' are neighbors if they can be written as $F=(T_1,\cdots,T_s)$ and $F'=(T_1',\cdots,T_s)$ where T_1 and T_1' are neighbors. The neighborhood relation defines a finite connected graph on all forests of equal size s with complexity not exceeding r. To each vertex F of this graph we assign a number $\phi(F)$. To find the global minimum of a function defined on a finite graph we apply the simulated annealing method (see, e.g., [24]). This method constructs some specified Markov process which takes values in the graph vertices and converges with high probability to the global minimum of the function. To avoid stalling at a local minimal point the process is allowed to pass with some small probability to a point F having greater value of $\phi(F)$ than current one. We propose a new modification of this method in which the output is the forest corresponding to the minimal value of a function $\phi(F)$ over all (randomly) visited points. Since simulated annealing performed involves random walks on a complicated graph consisting of trees as vertices, the algorithm was realized by means of the MSU supercomputer.

2.4. Machine Learning Methods

Let us describe two machine learning methods: random forests and stochastic gradient boosting. They will be used in Section 3.

We employ classification and regression trees (CART) as a base learning algorithm in RF and SGB because it showed good performance in a number of studies (see [18]). Classification tree T is a binary tree having the following structure. Any leaf of T contains either 1 or –1 and for any vertex P in T (including leaves) there exists a subset A_P of the explanatory variable space $\mathbb{X}$, such that the following properties hold:

1) $A_P = \mathbb{X}$, if P is the root of T.

2) If vertices P' and P'' are children of P, then

$A_{P'} \cup A_{P''} = A_P$ and $A_{P'} \cap A_{P''} = \emptyset$.

In particular, subsets corresponding to the leaves form the partition of $\mathbb{X}$. A classifier defined by a classification tree is introduced as follows. To obtain a prediction of Y given a certain value $x \in \mathbb{X}$ of the random vector X, one should go along the path which starts from the root and ends in some leaf turning at each parent vertex P to that child P' for which $A_{P'}$ contains x. At the end of the x-specific path, one gets either 1 or –1 which serves as a prediction of Y. Classification tree could be constructed via CART algorithm, see [18].

RF is a non-parametric method of estimating conditional probability $p = p(x)$. Its idea is to improve prediction power of CART tree by taking the average of these trees grown on many bootstrap samples, see [18, ch. 15]. The advantages of this method are low computational costs and the ability to extract relevant predictors when the number of irrelevant ones is large, see [25].

SGB is another non-parametric method of estimating conditional probability $p(x)$. SGB algorithm proceeds iteratively in such a way that, on each step, it builds a new estimate of $p(x)$ and a new classifier decreasing the number of misclassified cases from the previous step, see, e.g., [12].

Standard RF and SGB work poorly for unbalanced samples. One needs either to balance given datasets (as in [26]) before these methods are applied or use special modifications of RF and SGB. To avoid overfitting, permutation test is performed. A common problem of all machine learning methods is a complicated functional form of the final probability estimate $(\text{w.r.t. } x)\ \hat{p}(x,\xi)$. In genetic studies, one wants to pick up all relevant combinations of SNPs and risk factors, based on a biological pathway causing the disease. Therefore, the final estimate $\hat{p}(x,\xi)$ is to be analyzed. We describe one of possible methods for such analysis within RF framework called conditional variable importance measure (CVIM). One could determine CVIM for each predictor X_i in X and range all X_i in terms of this measure. Following [27], CVIM of predictor X_i given certain subvector Z_i of X is calculated as follows (supposing Z_i takes values $z_{i1},\cdots,z_{im(i)} \in \mathbf{X}^d$ for some $d \in \{1,\cdots,n\}$).

1) For each $k = 1,\cdots,m(i),$ permute randomly the elements of $A_{ik} = \{j : Z_i^j = z_{ik}\}$ to obtain a vector $\tau^{(k)} := \left(\tau_1^{(k)},\cdots,\tau_{M(k)}^{(k)}\right)$, where $M(k) = \# A_{ik}$. Consider a vector

$$(l_1,\cdots,l_N) := \left(\tau^{(1)},\cdots,\tau^{(m(i))}\right).$$

2) Let $B \in \mathbf{N}$. Generate bootstrap samples

$$b = 1,\cdots,B.\ \xi_b = \left(\left(X^{jb},Y^{jb}\right), j = 1,\cdots,N\right),$$

For each of these samples, construct a CART classifier $f_b(x,\xi_b)$ and calculate

$$CVIM_b = \frac{1}{|C_b|}\sum_{j\in C_b}\left(I\left\{Y^j = f_b\left(X^j,\xi_b\right)\right\} - I\left\{Y^j = f_b\left(X^{l_j},\xi_b\right)\right\}\right)$$

where $C_b = \left\{j \in \{1,\cdots,N\} : \left(X^j, Y^j\right) \notin \xi_b\right\}$.

3) Compute the final CVIM using the formula

$$CVIM = \frac{1}{B}\sum_{b=1}^{B} CVIM_b \quad .(29)$$

Any permutation $(l_1,\cdots,l_N)$ in the CVIM algorithm destroys dependence between X_i and (Y, Z_{-i}) where Z_{-i} consists of all components of X which are not in Z_i. At the same time it preserves initial empirical distribution of (X_i, Z_i) calculated for the sample x. The average loss of correctly classified Y is calculated, and if it is relatively large w.r.t. CVIM of other predictors, then X_i plays important role in classification and vice versa.

For instance, as Z_i ($i = 1,\cdots,n$) one can take all the components X_k ($k \neq i$) such that the hypothesis of the independence between X_k and X_i is not rejected at some significance level (e.g., 5%). CVIM-like algorithm could be used to range combinations of predictors w.r.t. the level of association to the disease. This will be published elsewhere.

3. APPLICATIONS: RISKS OF CHD AND MI

We employ here various statistical methods described above to analyze the influence of genetic and non-genetic (conventional) factors on risks of coronary heart disease and myocardial infarction using the data for 454 individuals (333 cases, 121 controls) and 333 individuals (165 cases, 168 controls) respectively. These data contain values of seven SNPs and four conventional risk factors. Namely, we consider glycoprotein Ia (GPIa), connexin- 37 (Cx37), plasminogen activator inhibitor type 1 (PAI-1), glycoprotein IIIa (GPIIIa), blood coagulation factor VII (FVII), coagulation factor XIII (FXIII) and interleukin-6 (IL-6) genes, as well as obesity (Ob), arterial hypertension (AH), smoking (Sm) and hypercholesterolemia (HC). The choice of these SNPs was based on biological considerations. For instance, to illustrate this choice we recall that connexin-37 (Cx37) is a protein that forms gapjunction channels between cells. The mechanism of the SNP in Cx37 gene influence on atherosclerosis development is not fully understood, but some clinical data suggest its importance for CHD development [28]. In the Russian population homozygous genotype can induce MI development, especially in individuals without CHD anamnesis [29].

The age of all individuals in case and control groups ranges from 35 to 55 years to reduce its influence on the risk analysis. For each of considered methods, we use K-fold cross-validation with K = 6. As shown in [16], the standard choice of partition number of cross-validation from 6 to 10 does not change the EPE significantly. We take K = 6 as the sample sizes do not exceed 500. The MSU supercomputer "Chebyshev" was involved to perform computations. As shown below, all applied methods permit to choose appropriate models having EPE for CHD dataset less than 0.25. Thus predictions constructed have significant predictive power. Note that, e.g., in [30] the interplay between genotype and MI development was also studied, with estimated prediction errors 0.30 - 0.40.

3.1. MDR and MDRIR Method

Coronary heart disease. **Table 1** contains EPEs of the most significant combinations of predictors obtained by MDR analysis of coronary heart disease data. Note that we write the gene meaning the corresponding SNP. To estimate the empirical c.d.f. of the prediction error when the disease is not linked with explanatory variables, we used the permutation test. Namely, 100 random uniform permutations of variables $Y^1,\dots,Y^N$ were generated. For any permutation $(Y_b^1,\dots,Y_b^N)$ we constructed a sample

$$\bar{\xi}_b = \left((X^1,Y_b^1),\dots,(X^N,Y_b^N)\right)$$

and applied the same analysis to this simulated sample, here $b=1,\dots,100$. In these 100 simulations the corresponding empirical prediction error was not less than 0.42. Thus the Monte Carlo p-value of all three combinations was less than 0.01 (since their EPEs were much less than 0.42), which is usually considered as a good performance.

Table 1 contains also the results of MDRIR method, which are similar to results of MDR method. However, MDRIR method allows to identify additional combinations (listed in this table) with EPE around 0.24. It follows from the same table that hypertension and hypercholesterolemia are the most important non-genetic risk factors. Indeed, these two factors appear in each of 6 combinations.

Table 1. The most significant combinations obtained by MDR and MDRIR analysis for CHD and MI data.

Disease	Method	Factors	EPE
CHD	MDR	GPIa, FXIII, AH, HC	0.231
		Cx37, AH, HC	0.238
		GPIa,Cx37, AH, HC	0.241
	MDRIR	FXIII, FVII, AH, HC	0.240
		FXIII, AH, HC	0.242
		GPIa, Cx37, AH, HC	0.247
MI	MDR	GPIIIa, FXIII, Cx37, AH	0.343
		GPIIIa, FXIII, FVII, Cx37	0.347
		Cx37, Sm	0.356
	MDRIR	Cx37, Sm	0.351
		GPIIIa, Cx37, Sm	0.353
		GPIIIa, Cx37, Sm, HC	0.355

To perform a more precise analysis of influence of SNPs on CHD provoking we analyzed gene-gene interactions. We used two different strategies. Namely, we applied MDR method to a subgroup of individuals who were not subject to any of the non-genetic risk factors, i.e. to non-smokers without obesity and without hypercholesterolemia, 51 cases and 97 controls (risk-free sample). Another strategy was to apply MDR method to the whole sample, but to take into account only genetic factors rather than all factors. **Table 2** contains the most significant combinations of SNPs and their EPEs.

Thus based on coronary heart disease data with the help of Tables 1 and 2 we can make the following conclusions. Combination of two SNPs (in GPIa and Cx37) and two non-genetic factors (hypertension and hypercholesterolemia) has the biggest influence on CHD. Also FXIII gives additional predictive power if AH and HC are taken into account.

It turns out that both methods yield similar results. Combination of SNPs in GPIa and Cx37 has the biggest influence on CHD. EPE is about 0.28 - 0.34, and smaller error corresponds to a risk-free sample. Moreover, it follows from Tables 1 and 2 that EPE dropped significantly after additional non-genetic factors were taken into account (the error is 0.247 if additional non-genetic factors are taken into account and 0.343 if not).

Myocardial infarction. EPEs of the most significant combinations obtained by MDR analysis of MI data are presented in **Table 1**. For all 100 simulations of $\overline{\xi}_b$ when the disease was not linked with risk factors, EPE was larger than 0.38. Monte Carlo p-value of all combinations was less than 0.01. MDRIR analysis of the same dataset gave a clearer picture (see the same table), as the pair (Cx37, Sm) appears in all three combinations with the least estimated prediction error.

Apparently, combination of smoking and SNP in Cx37 is the most significant. These two factors appear in all combinations in **Table 1** concerning MI and MDRIR. Involving any additional factors only increases EPE.

Table 2. Comparison of the most significant SNP combinations obtained by two different ways of MDR analysis of CHD data.

Method	Factors	EPE
MDR on individuals not subject to risk factors	GPIa, Cx37	0.281
MDR on the whole group concerning only genetic data	GPIa, Cx37	0.343

The explicit form of the prediction algorithm based on Cx37 and Sm shows that these factors exhibit interaction. Smoking as well as homozygote for recessive allele of the SNP in Cx37 provokes the disease. However wildtype allele can protect from consequences of smoking. Namely, the combination of smoking and Cx37 wildtype is protective, i.e. the value of prediction algorithm of this combination is –1.

3.2. Ternary Logic Regression

We performed several research procedures for CHD and MI data, with different restrictions imposed on the statistical model. Set

$$(X_1,\cdots,X_n)=(Z_1,\cdots,Z_m,R_1,\cdots,R_k)$$

where a vector $Z=(Z_1,\cdots,Z_m)$ stands for SNP values (in PAI-1, GPIa, GPIIIa, FXIII, FVII, IL-6, Cx37 respectively) and $R=(R_1,\cdots,R_k)$ denotes non-genetic risk factors (Ob, AH, Sm, HC), $k=4.\ m=7,$

We considered four different models in order to analyze both total influence of genetic and non-genetic factors and losses in predictive force appearing when some factors were excluded. In our applications we took $s=3$ as search over larger forests for samples with modest sizes could have given very complicated and unreliable results.

Model 1. Define the class M (see Section 2.3) consisting of the functions h having a form

$$h(Z,R)=\beta_0+\sum_{v=1}^{s}\beta_v T_v(Z_1,\cdots,Z_m)+\sum_{v=1}^{k}\beta_{s+v}R_v$$

where $\beta_v \in \mathbf{R}$ and T_v are polynomials identified with trees. In other words we require that non-genetic factors are present only in trees consisting of one variable.

Model 2. Now we assume that any function $h \in M$ has the representation

$$h(Z,R)=\beta_0+\sum_{v=1}^{s}\beta_v T_v(Z_1,\cdots,Z_m,R_1,\cdots,R_k)$$

where $\beta_v \in \mathbf{R}$ and T_v are polynomials identified with trees. Thus we allow the interaction of genes and nongenetic factors in order to find significant gene-environment interactions. However we impose additional restrictions to avoid too complex combinations of nongenetic risk factors. We do not tackle here effects of interactions where several non-genetic factors are involved. Namely, we consider only the trees satisfying the following two conditions.

1) If there is a leaf containing non-genetic factor variable then the root of that leaf contains product operator.

2) Moreover, another branch growing from the same root is also a leaf and contains a genetic (SNP) variable.

Models 3 and 4 have additional restrictions that polynomials T_v $(v=1,\cdots,s)$ in (30) depend only on nongenetic factors and only on SNPs respectively. These models are considered to compare their results with ones obtained with all information taken into account, in order to demonstrate the importance of genetic (resp. non-genetic) data for risk analysis.

Coronary heart disease. The obtained results are provided in **Table 3**.

EPE in Model 1 for CHD was only 0.19. For the same model we performed also fast simulated annealing search of the optimal forest which was much more time-efficient, and a reasonable error of 0.23 was obtained. Model 3 application showed that non-genetic factors play an important role in CHD genesis, as classification based on non-genetic factors only gave the error less than 0.23, while usage of SNPs only (Model 4) let the error grow to 0.34.

Model 1 gave the minimal EPE. For the optimal forest $(T_1,\cdots,R_4)$ the function $\hat{h}(Z,R)$ given before formula (28) with $S=\{1,\cdots,N\}$ is provided by

$$\begin{aligned}&-0.597T_1-0.354T_2+0.521T_3-0.444R_1\\&+1.311R_2-0.146R_3+2.331R_4-0.226\end{aligned} \qquad (31)$$

where

$$\begin{aligned}T_1&=(Z_3Z_4+Z_6Z_7+(Z_2)^2+Z_3Z_7)(Z_1)^2Z_3Z_7,\\T_2&=Z_1(Z_3)^2(Z_6Z_7+Z_2(Z_4)^2Z_7),\\T_3&=Z_2+2Z_2(Z_6)^2Z_7\end{aligned}$$

with sums and products modulo 3.

The non-genetic factors 2 and 4 (i.e. AH and HC) are the most influential since the coefficients at them are the greatest ones (1.311 and 2.331). As is shown above, MDR yielded the same conclusion. If the gene-environment interactions were allowed (Model 2), no considerable increase in predictive power has been detected. However we list the pairs of SNPs and non-genetic factors present in the best forest: Z_7 and R_2, Z_7 and R_1, Z_7 and R_4, Z_5 and R_1 We see that SNP in Cx37 is of substantial importance as it appears in combination with all risk factors except for smoking.

As formula (31) is hard to interpret, we select the most significant SNPs via a variant of permutation test. Consider a random rearrangement of the column with first SNP in CHD dataset. Calculate the EPE using these new simulated data and the same function $\hat{h}$ as before. The analogous procedure is done for other columns (containing the values of other SNPs) and the errors found are given in **Table 4** (recall that the EPE equals 0.19 if no permutation is done).

It is seen that the error increases considerably when the values of GPIa and Cx37 are permuted. The statement that they are the main sources of risk agrees with what was obtained above by MDR method.

Myocardial infarction. For the MI dataset, under the same notation that above, the results obtained for our four models are given in **Table 3**. To comment them we should first note that non-genetic risk factors play slightly less important role compared with CHD risk: if they are used without genetic information, the error increases by 0.09, see Models 1 and 3 (while the same increase for CHD was 0.03). The function $\hat{h}(Z,R)$ defined before (28) with $S=\{1,\cdots,N\}$ equals

$$-1.144T_1+0.914T_2-0.45T_3-0.285R_1$$
$$-0.675R_2+0.828R_3-0.350R_4-0.055$$

where $T_1=Z_1Z_3(Z_5)^2$, $T_2=Z_7$, $T_3=Z_3+Z_4+Z_6+Z_7$.

Thus the first tree has the greatest weight (coefficient equals –1.144), the second tree (i.e. SNP in Cx37) is on the second place, and non-genetic factors are less important.

Table 3. Results of TLR.

Model	1	2	3	4
EPE for CHD	0.190	0.204	0.228	0.340
EPE for MI	0.305	0.331	0.391	0.365

Table 4. The SNP significance test for CHD in Model 1.

SNP permuted	EPE
GPIa	0.263
Cx37	0.260
IL-6	0.226
PAI-1	0.212
GPIIIa	0.208
FXIII	0.202
FVII	0.190

As for CHD we performed a permutation test to compare the significance of different SNPs. Its results are presented in **Table 5**.

As seen from this table, the elimination of Cx37 SNP leads to a noticeable increase in the EPE. This fact agrees with results obtained by MDR analysis of the same dataset.

Table 5. The SNP significance test for MI in Model 1.

SNP permuted	EPE
Cx37	0.444
GPIIIa	0.353
IL-6	0.340
FXIII	0.328
FVII	0.324
PAI-1	0.319
GPIa	0.305

3.3. Results Obtained by RF and SGB Methods

Since machine learning methods give too complicated estimates of the dependence structure between Y and X, we have two natural ways to compare them with our methods. Namely, these are the prediction error and the final significance of each predictor. The given datasets were unbalanced w.r.t. response variable and we first applied the resampling technique to them. That means enlargement of the smaller of two groups case-control in the sample by additional bootstrap observations until the final proportion case:control is 1:1. Note that due to the resampling techniques the following effect arises: some observations in small groups (case or control) appear in the new sample more frequently than other ones. Therefore, we took the average over 1000 iterations.

Coronary heart disease. Results of RF and SGB methods are given in **Table 6**. It shows that RF and SGB methods gave statistically reliable results (EPE in the permutation test is close to 50%). Moreover, additional SNP information improved predicting ability by 13% (SGB). It seems that SGB method is fitted better to CHD data than RF.

To compute CVIM for each X_i, we constructed a vector Z_i as follows. Let Z_i contain all predictors X_j, $j \neq i$, for which chi-square criteria rejected independence hypothesis between X_j and X_i at 5% significance level. **Table 7** shows that the most relevant predictors for CHD are AH, HC and Cx37.

Table 6. EPE/EPE in permutation test calculated via crossvalidation for CHD and MI datasets with employment of RF and SGB methods.

Disease	Data	RF	SGB
CHD	with SNP	0.200/0.454	0.134/0.473
	without SNP	0.230/0.510	0.261/0.503
MI	with SNP	0.360/0.497	0.399/0.530
	without SNP	0.473/0.527	0.482/0.562

Table 7. Predictors ranged in terms of their CVIM for CHD and MI dataset.

CHD		MI	
AH	8.90	Cx37	7.50
HC	5.30	Sm	2.00
Cx37	5.10	AH	1.86
Ob	0.56	GPIIIa	0.03
FXIII	0.53	FVII	0.02
Sm	0.11	FXIII	≈0
GPIa	0.10	HC	≈0
FVII	0.07	GPIa	≈0
PAI-1	0.03	Ob	≈0
GPIIIa	0.02	IL-6	≈0
IL-6	0.01	PAI-1	≈0

Myocardial infarction. Results of RF and SGB methods are given in **Table 6**. It shows that RF and SGB methods gave statistically reliable estimates (EPE in the permutation test is close to 50%). Moreover, additional SNP information improved predicting ability by 11% (RF method).

CVIM was calculated according to (29) and is given in **Table 7**. Thus, the most relevant predictors for MI are Cx37, Sm and AH.

4. CONCLUSIONS AND FINAL REMARKS

In the current study we developed important statistical methods concerning the analysis of multidimensional genetic data. Namely, we proposed the MDR with independent rule and ternary logic regression with a new version of simulated annealing. We compared them with several popular methods which appeared during the last decade. It is worth to emphasize that all considered methods yielded similar qualitative results for dataset under study.

Let us briefly summarize the main results obtained. The analysis of CHD dataset showed that two non-genetic risk factors out of four considered (AH and HC) had a strong connection with the disease risk (the error of classification based on non-genetic factors only is 0.25 - 0.26 with p-value less than 0.01). Also, the classification based on SNPs only gave an error of 0.28 which is close to one obtained by means of non-genetic predictors. Moreover, the most influential SNPs were in genes Cx37 and GPIa (FXIII also entered the analysis only when AH and HC were present). EPE decreased to 0.13 when both SNP information and non-genetic risk factors were taken into account and SGB was employed. Note that excludeing any of the 5 remaining SNPs (all except for two most influential) from data increased the error by 0.01 - 0.02 approximately. So, while the most influential data were responsible for the situation within a large part of population, there were smaller parts where other SNPs came to effect and provided a more efficient prognosis ("small subgroups effect"). The significance of SNP in GPIa and FXIII genes was observed in our work. AH and HC influence the disease risk by affecting the vascular wall, while GPIa and FXIII may improve prognosis accuracy because they introduce haemostatic aspect into analysis.

The MI dataset gave the following results. The most significant factors of MI risk were the SNP in Cx37 (more precisely, homozygous for recessive allele) and smoking with a considerable gene-environment interacttion present. The smallest EPE of methods applied was 0.33 - 0.35 (with p-value less than 0.01). The classification based on non-genetic factors only yielded a greater error of 0.42. Thus genetic data improved the prognosis quality noticeably. While two factors were important, other SNPs considered actually did not improve the prognosis essentially, i.e. no small groups effect was observed.

While CHD data used in the study permitted to specify the most important predictors with EPE about 0.13, the MI data lead to less exact prognoses. Perhaps this complex disease requires a more detailed description of individual's genetic characteristics and environmental factors.

The conclusions given above are based on several complementary methods of modern statistical analysis. These new data mining methods allow to analyze other datasets as well. The study can be continued and the medical conclusions need to be replicated with larger datasets, in particular, involving new SNP data.

5. ACKNOWLEDGEMENTS

The work is partially supported by RFBR grant (project 10-01-00397a).

REFERENCES

1. Y. Fujikoshi, R. Shimizu and V. V. Ulyanov, "Multivariate Statistics: High-Dimensional and Large-Sample Approximations," Wiley, Hoboken, 2010.

2. S. Szymczak, J. Biernacka, H. Cordell, O. GonzálezRecio, I. König, H. Zhang and Y. Sun, "Machine Learning in Genome-Wide Association Studies," Genetic Epidemiology, Vol. 33, No. S1, 2009, pp. 51-57.

3. D. Brinza, M. Schultz, G. Tesler and V. Bafna, "RAPID Detection of Gene-Gene Interaction in Genome-Wide Association Studies", Bioinformatics, Vol. 26, No. 22, 2010, pp. 2856-2862.

4. K. Wang, S. P. Dickson, C. A. Stolle, I. D. Krantz, D. B. Goldstein and H. Hakonarson, "Interpretation of Association Signals and Identification of Causal Variants from Genome-Wide Association Studies," The American Journal of Human Genetics, Vol. 86, No. 5, 2010, pp. 730-742.

5. Y. Liang and A. Kelemen. "Statistical Advances and Challenges for Analyzing Correlated High Dimensional SNP Data in Genomic Study for Complex Diseases," Statistics Surveys, Vol. 2, No. 1, 2008, pp. 43-60.

6. H. Schwender and I. Ruczinski, "Testing SNPs and Sets of SNPs for Importance in Association Studies," Biostatistics, Vol. 12, No. 1, 2011, pp. 18-32.

7. M. Ritchie, L. Hahn, N. Roodi, R. Bailey, W. Dupont, F. Parl and J. Moore, "Multifactor-Dimensionality Reduction Reveals High-Order Interactions among EstrogenMetabolism Genes in Sporadic Breast

Cancer," The American Journal of Human Genetics, Vol. 69, No. 1, 2001, pp. 138-147.

8. D. Velez, B. White, A. Motsinger, W. Bush, M. Ritchie, S. Williams and J. Moore, "A Balanced Accuracy Function for Epistasis Modeling in Imbalanced Datasets Using Multifactor Dimensionality Reduction," Genetic Epidemiology, Vol. 31, No. 4, 2007, pp. 306-315.

9. I. Ruczinski, C. Kooperberg and M. LeBlanc, "Logic Regression," Journal of Computational and Graphical Statiststics, Vol. 12, No. 3, 2003, pp. 475-511.

10. H. Schwender and K. Ickstadt, "Identification of SNP Interactions Using Logic Regression," Biostatistics, Vol. 9, No. 1, 2008, pp. 187-198.

11. L. Breiman, "Random Forests," Machine Learning, Vol. 45, No. 1, 2001, pp. 5-32.

12. J. Friedman, "Stochastic Gradient Boosting," Computational Statistics & Data analysis, Vol. 38, No. 4, 2002, pp. 367-378.

13. X. Wan, C. Yang, Q. Yang, H. Xue, N. Tang and W. Yu, "Mega SNP Hunter: A Learning Approach to Detect Disease Predisposition SNPs and High Level Interactions in Genome Wide Association Study," BMC Bioinformatics, Vol. 10, 2009, p. 13.

14. A. Bulinski, O. Butkovsky, A. Shashkin, P. Yaskov, M. Atroshchenko and A. Khaplanov, "Statistical Methods of SNPs Analysis," Technical Report, 2010, pp. 1-159 (in Russian).

15. G. Bradley-Smith, S. Hope, H. V. Firth and J. A. Hurst, "Oxford Handbook of Genetics," Oxford University Press, New York, 2010.

16. S. Winham, A. Slater and A. Motsinger-Reif, "A Comparison of Internal Validation Techniques for Multifactor Dimensionality Reduction," BMC Bioinformatics, Vol. 11, 2010, p. 394.

17. A. Arlot and A. Celisse, "A Survey of Cross-Validation Procedures for Model Selection," Statistics Surveys, Vol. 4, No. 1, 2010, pp. 40-79.

18. T. Hastie, R. Tibshirani and J. Friedman, "The Elements of Statistical Learning: Data Mining, Inference, and Prediction," 2nd Edition, Springer, New York, 2009.

19. R. L. Taylor and T.-C. Hu, "Strong Laws of Large Numbers for Arrays of Rowwise Independent Random Elements," International Journal of Mathematics and Mathematical Sciences, Vol. 10, No. 4, 1987, pp. 805-814.

20. E. Lehmann and J. Romano, "Testing Statistical Hypotheses," Springer, New York, 2005.

21. P. Golland, F. Liang, S. Mukherjee and D. Panchenko, "Permutation Tests for Classification," Lecture Notes in Computer Science, Vol. 3559, 2005, pp. 501-515.

22. J. Park, "Independent Rule in Classification of Multivariate Binary Data," Journal of Multivariate Analysis, Vol. 100, No. 10, 2009, pp. 2270-2286.

23. S. Lee, Y. Chung, R. Elston, Y. Kim and T. Park, "LogLinear Model-Based Multifactor Dimensionality Reduction Method to Detect Gene-gene Interactions," Bioinformatics, Vol. 23, No. 19, 2007, pp. 2589-2595.

24. A. Nikolaev and S. Jacobson, "Simulated Annealing," In: M. Gendreau and J.-Y. Potvin, Eds., Handbook of Metaheuristics, Springer, New York, 2010, pp. 1-39.

25. G. Biau, "Analysis of a Random Forests Model," LSTA, LPMA, Paris, 2010.

26. N. Chawla, "Data Mining for Imbalanced Datasets: An Overview," In: O. Maimon and L. Rokach, Eds., Data Mining and Knowledge Discovery Handbook, Springer, New York, 2010, pp. 875-886.

27. C. Strobl, A. Boulesteix, T. Kneib, T. Augustin and A. Zeileis, "Conditional Variable Importance for Random Forests," BMC Bioinformatics, Vol. 9, 2008, p. 307

28. A. Hirashiki, Y. Yamada, Y. Murase, Y. Suzuki, H. Kataoka, Y. Morimoto, T. Tajika, T. Murohara and M. Yokota, "Association of Gene Polymorphisms with Coronary Artery Disease in Lowor High-Risk Subjects Defined by Conventional Risk Factors," Journal of the American College of Cardiology, Vol. 42, No. 8, 2003, pp. 1429-1437. A. Balatskiy, E. Andreenko and L. Samokhodskaya, "The Connexin37 Polymorphism as a New Risk Factor of MI Development," Siberian Medical Journal, Vol. 25, No. 2, 2010, pp. 64-65 (in Russian).

29. C. Coffey, P. Hebert, M. Ritchie, H. Krumholz, J. Gaziano, P. Ridker, N. Brown, D. Vaughan and J. Moore, "An Application of Conditional Logistic Regression and Multifactor Dimensionality Reduction for Detecting GeneGene Interactions on Risk of Myocardial Infarction: The Importance of Model Validation," BMC Bioinformatics, Vol. 5, 2004, p. 49.

Chapter 9

A Multivariate Statistical Sampling Technique to Enhance Far Quantile Estimates of Arbitrary Responses

Christian Caillat [1], **Eric Carman** [2]

[1]*Emerging Memory Group, Micron Technology Belgium, Leuven, 3001, Belgium*

[2]*Emerging Memory Group, Micron Technology, San-Jose (CA), 95134-5134, USA*

ABSTRACT

We propose a statistical sampling method, called eXtreme Event Sampling (XES), to compute far quantiles of arbitrary responses of multiple independent random parameters more accurately and efficiently than with Classical Monte-Carlo (CMC). Based on the selective over-sampling of events of low probability of occurrence, the method enables the study of multiple responses at a time, unlike the classical Importance Sampling (IS) techniques, expected to be the best-performing when tuned to a single given response. Though more generic than IS, XES still shows large gains over CMC in both accuracy and sampling efficiency, even for a large number of parameters (up to 27 tested). This article presents the detailed theoretical aspects of XES and an empirical study demonstrating its efficiency in various cases of responses (linear or non-linear), distributions (normal and non-normal) and number of parameters. If the primary target application is the design of semiconductor memory circuits, we believe that the flexibility of the method potentially makes it attractive in other contexts showing similar constraints.

Keywords: Importance Sampling, Monte-Carlo, Statistical Variations, Low Probability Events

1. INTRODUCTION

Strategies for quick and accurate computation of Cumulative Distribution Functions (CDF) and more especially far quantiles is a topic common to a large variety of domains, such as finance, communications, insurance or nuclear physics[1][2], which led researchers to propose a collection of solutions more efficient than the "brute force" of Classical (or "Crude") Monte-Carlo (CMC). The family of Importance Sampling (IS) strategies is certainly among the most efficient ones to achieve that[3], at least theoretically, since the ideal sampling scheme requires knowing the response (or even better: its density of probability) in order to adjust the sampling to it[1][3]. Indeed, blindly applying an IS strategy such as *scaling* or *translation*[3][4], exposes to the risk of getting an increasing variance instead of a decreasing variance of quantile estimates[5].

In the context of microelectronics circuit design, and more particularly semiconductor memories, of primary interest to us, there has been a growing interest in these computation-effective solutions over the past years[6 -9].

The fundamental reason lies in the growing impact of the various sources of statistical fluctuations of unit devices composing the memory cells as technologies progress [10-13] and in the rapid growth of memory capacity, now already entering the 10-100Gb era for standalone memories[14]. In the case of MOSFET transistors, for example, the threshold voltage fluctuation is a well known effect originating from multiple physical sources of randomness[13].

If not taken into account properly, these fluctuations are likely to degrade the overall design performance and ultimately compromise the functionality itself, if the failure rate exceeds a critical value (depending on the redundancy scheme used, in the case of memories[15]).

Classically, to assess memory design functionality towards cell fluctuations, CMC is performed and various critical signals of the circuit are analyzed (voltages, currents, delays, etc.). Since CMC is CPU intensive and practically ineffective in predicting far quantiles, it is not suitable for Giga-bit arrays. An ideal replacement strategy would not only show significant gain over CMC but be versatile enough to analyze multiple arbitrary signals in a single set of simulation runs, which does not fit well the classical IS strategies, where the sampling is adaptive to a response. Indeed, in the case of multiple responses, applying an IS strategy tuned for a given response to another one, the same above-mentioned risk of a diverging variance exists.

For this reason, we recently proposed[16][17] a more generic sampling strategy based on the enhancement of rare events – i.e. with a low probability of occurrence in the parameter space – suitable to any response and that does not require knowing the responses or even deciding beforehand which ones will be analyzed. The proposed method is called eXtreme Event Sampling (XES), as it favors the rare events that are significantly deviating from the nominal parameter conditions for which the circuit is expected to be in its ideal operation region. More specifically, a refined sampling is applied in selected probability ranges, which extends the CDF calculation of a response to far quantiles. In that sense, since XES enhances specific categories of events, it can be considered as an IS scheme and actually shows similar formulas, as we will expose in this article.

In a first section, we will detail the theoretical background supporting the construction of the sampling method, including the transformation allowing non-normal parameter distributions to be taken into account. In a second section, we will illustrate the effectiveness of XES through empirical studies for various responses, distributions and number of parameters and finally provide a benchmark of XES to CMC.

Definitions:

X_k^i : a vector of k i.i.d. random variables at iteration i

$R(X_k^i)$: response (scalar) of a vector X_k^i

f(.): probability density function (PDF)

F(.): cumulative distribution function (CDF)

$F^{-1}(.)$: inverse CDF

I(T): indicator function yielding 0 if T is false, 1 otherwise

$N(\mu,\sigma^2)$: normal distribution of mean μ and variance σ^2

ξ_p: p-quantile of a distribution = $F^{-1}(p)$

$\tilde{\xi}_p$: estimator of ξ_p

$W_i(.)$: i^{th} statistical weight of an IS run

N: total number of samples in MC or IS run

$\Phi(.)$: standard normal CDF

z: radius of a k-sphere

$n(z)$: function yielding a number of samples per sphere

All random variables are assumed independent

2. CONSTRUCTION OF THE SAMPLING METHOD

2.1. The Multivariate Normal Distribution

The lack of information on the responses does not prevent calculating a probability of occurrence of events in a multivariate problem. The idea of XES is to do so in order to over-sample rare events and thus enhance the CDF calculation for far quantiles as compared to a CMC sampling. In order to identify such rare events, we first have to establish the joint probability of a set of random variables (x_k).

With the assumptions that all input variables are normally distributed and independent, the multivariate joint Probability Density Function (PDF) of a set x_k of such variables ($x_k \sim N(\mu_k,\sigma_k^2)$) reads[18]:

$$f(x_k) = \frac{1}{(2\pi)^{k/2} \cdot \prod_k \sigma_k} \cdot e^{-\frac{1}{2}\sum_k \frac{(x_k - \mu_k)^2}{\sigma_k^2}} \quad (1)$$

Iso-density curves for such a PDF are defined by assuming a constant exponential term as in (2), thus forming a family of parameterized ellipsoids[18] (see illustration for k=2 in Figure 1). The constant parameters z, μ_k and $\sigma_k$fully characterize the ellipsoid.

$$\sum_k \frac{(x_k - \mu_k)^2}{\sigma_k^2} = z^2 \quad (2)$$

Applying a change of variable $X_k = (x_k-\mu_k)/\sigma_k$, (2) now represents a hyper-sphere with respect to X_k in a space of dimension k (or *k-sphere*), and z is its radius. In the following, for the convenience of notations, we will always consider variables normalized to the standard N(0,1) distribution, unless otherwise specified. The choice of hyper-spherical coordinates to represent the samples is then the most natural one: the k Cartesian coordinates are represented by a radius z and (k-1) angles θ_i[19]:

$$\begin{aligned}
X_1 &= z \cdot \cos(\theta_1) \\
X_2 &= z \cdot \sin(\theta_1) \cdot \cos(\theta_2) \\
X_3 &= z \cdot \sin(\theta_1) \cdot \sin(\theta_2) \cdot \cos(\theta_3) \\
&\vdots \\
X_{k-1} &= z \cdot \sin(\theta_1) \cdots \sin(\theta_{k-2}) \cdot \cos(\theta_{k-1}) \\
X_k &= z \cdot \sin(\theta_1) \cdots \sin(\theta_{k-2}) \cdot \sin(\theta_{k-1})
\end{aligned} \quad (3)$$

Using such hyper-spherical coordinates (3) and normalized variables (or μ_k=0, σ_k=1), the PDF (1) now simply reads:

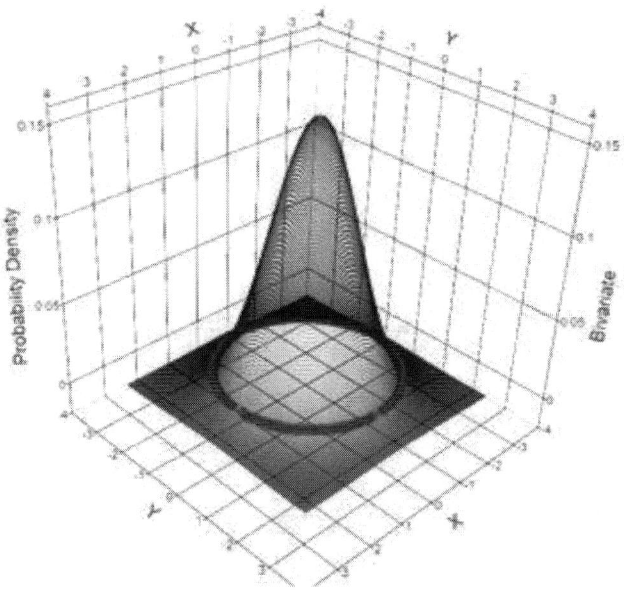

Figure 1. Example of bivariate probability density function with an iso-density ellipse highlighted. X, Y ~ N(0,1). X/Y scales in standard sigma

$$f(z,\theta_i)=\frac{1}{(2\pi)^{k/2}}\cdot e^{-\frac{z^2}{2}} \tag{4}$$

As expected, since the hyper-spheres represent iso-density curves (radial symmetry of the density of probability), (4) shows no dependence to the (k-1) angles θ_i of the system.

As a consequence, calculating the CumulativeDistribution Function (CDF) from (4) using these coordinates can be done regardless of the angular position and with a single variable of integration: z. It also implies that a given k-sphere is not only an iso-density surface but also an iso-probability surface. Finally, since the variable z represents the distance from the center point of the parameter space (z=0 means all variables are at their average value), low probability events of interest for XES corresponds to high z values – this aspect is further commented in section 2.5, especially as regards the case of circuit analysis.

In the proposed technique, we take advantage of these properties to calculate the CDF of responses of multiple variables, using randomly chosen samples on concentric hyper-spheres and a modulation of the sampling versus the parameter z to enhance rare events in selected ranges of probability of occurrence.

We will now describe how to achieve such a sampling plan in practice and how to calculate the CDF of arbitrary responses.

2.2. Marginal PDF on k-Spheres

Let's first establish the natural density of samples on a given k-sphere, by calculating the following marginal PDF (p_z) of events for a k-sphere of radius z:

$$p_z = \int_{\theta_i} f(z,\theta_i) \cdot dV = \frac{1}{(2\pi)^{k/2}} \cdot e^{-\frac{z^2}{2}} \cdot \int_{\theta_i} dV \tag{5}$$

where dV is the volume element (Jacobian) of the k-sphere. In the hyper-spherical coordinates, dV reads[19]:

$$dV = z^{k-1} \cdot dz \cdot \prod_{i=0}^{k-2} \sin^i(\theta_{k-i-1}) \cdot d\theta_{k-i-1} \tag{6}$$

Hence, we can rewrite the integral in (5):

$$\begin{aligned} \int_{\theta_i} dV &= z^{k-1} \cdot dz \cdot \int_{\theta_i} \prod_{i=0}^{k-2} \sin^i(\theta_{k-i-1}) \cdot d\theta_{k-i-1} \\ &= z^{k-1} \cdot dz \cdot S_k \end{aligned} \tag{7}$$

with S_k being, by definition, the area of the unit k-sphere[20]:

$$S_k = \frac{2\pi^{k/2}}{\Gamma(k/2)}, \Gamma(t) = \int_0^{\infty} t^{z-1} \cdot e^{-t} \cdot dt \tag{8}$$

As a result, p_z reads:

$$p_z = \int_{\theta_i} f(z,\theta_i) \cdot dV = \frac{2^{1-k/2}}{\Gamma(k/2)} \cdot z^{k-1} \cdot e^{-\frac{z^2}{2}} \cdot dz \tag{9}$$

then, calculating the corresponding CDF from (9) requires integrating p_z with z varying from 0 to +∞.

2.3. Empirical CDF Calculation

Let's now introduce a function *n*(z) yielding a number of samples *n* for a given k-sphere radius z. Since the natural density of samples for a given sphere z is defined by (9), the corrected marginal PDF for a such a sampling plan equals $p_z/n(z)$. Taking into account the above and similarly to the classical Monte-Carlo technique[21], we can then compute the empirical probability F(ξ) for a response $R(X_k)$ to be lower than ξ from the following discrete sum (in other words: the discrete CDF of R):

$$F(\xi) = P(R(X_k) < \xi)$$

$$= \sum_{i=1}^{N} I\left(R\left(X_k^i\right) < \xi\right) \cdot \frac{2^{1-k/2}}{n(z_i) \cdot \Gamma(k/2)} \cdot z_i^{k-1} \cdot e^{-\frac{z_i^2}{2}} \cdot dz \qquad (10)$$

for a given series of radiuses z_i and a constant stepping dz, and with N being the total number of samples:

$$N = \sum_{z=z\,\min}^{z\,\max} n(z) \qquad (11)$$

F(ξ) can also be written:

$$F(\xi) = P(R(X_k) < \xi) = \sum_{i=1}^{N} I\left(R\left(X_k^i\right) < \xi\right) \cdot W(z_i)$$

$$W(z_i) = \frac{2^{1-k/2}}{n(z_i) \cdot \Gamma(k/2)} \cdot z_i^{k-1} \cdot e^{-\frac{z_i^2}{2}} \cdot dz \qquad (12)$$

Such a formulation is similar to Importance Sampling techniques[3], $W(z_i)$ representing the statistical weight associated to samples. In practice, the integration variable z is taken on a constant grid (dz constant), which also makes the integration method somewhat similar to a quasi-Monte-Carlo approach[21], but with a grid representing here equi-probable occurrences.

Theoretically, the sum of weights thus defined (12) should exactly equal one. In practice, however, discrete summation and numerical error lead to significant deviations from this ideal situation, which suggests performing a normalization of weights to their average, as often reported in the literature[3][22]:

$$F(\xi) = P(R(X_k) < \xi) = \frac{1}{N} \sum_{i=1}^{N} I\left(R\left(X_k^i\right) < \xi\right) \cdot \frac{W(z_i)}{\overline{W}}$$

$$\overline{W} = \frac{1}{N} \sum_{i=1}^{N} W(z_i) \qquad (13)$$

This last formulation (13) is the selected method for all CDF calculations presented below.

Finally, it is easy to verify that if $n(z_i)$ is chosen equal to p_z (9), then $W(z_i)=1$ hence (13) becomes a CMC sampling.

2.4. Uniform Sampling on k-Spheres

We implicitly assumed in the above reasoning that the *n* samples taken on a k-sphere of radius z were equi-probable. In order to respect this condition, we chose to use the following method, introduced by Marsaglia[23], to draw samples over a k-sphere with a uniform spatial distribution:

$$x_i \sim \mathcal{N}(0,1), S=\sum_{i=1}^{k} x_i^2$$

$$X_i = \frac{x_i}{S^{1/2}}, i \in [1,k] \tag{14}$$

where (X_i) are the k Cartesian coordinates of the samples.

The k normally distributed samples are drawn, for example, by using the inverse normal CDF $\Phi^{-1}(u)$, with u being uniformly distributed over$[0,1[$. Alternatively, a low-discrepancy sequence or a stratification technique (such as Latin Hypercube Sampling – LHS)[21][24] can be used to draw the required u_k samples from the hyper-cube$[0,1[^k$. We briefly verified that the evenness of sample distribution over k-spheres was indeed improved using a LHS strategy (not illustrated here). In the rest of the article, we will assume that simple uniform random samples u_i and the inverse normal function $\Phi^{-1}(u_i)$ are used to generate the above X_i coordinates.

2.5. Sampling Plan: Defining *n*(z)

The last part of the method consists in choosing a number of samples per k-sphere for each radius z.

The primary goal of XES being to favor rare events in a selected range of occurrence probability, we first have to compute the CDF of events from the joint PDF (9). The PDF for various k values are illustrated in Figure 2, and (1-CDF) is shown in Figure 3.

We stress again here that this represents the probability of a set of events belonging to a given k-sphere to occur, regardless of the actual values of the response, since the response is assumed to be unknown at this stage.

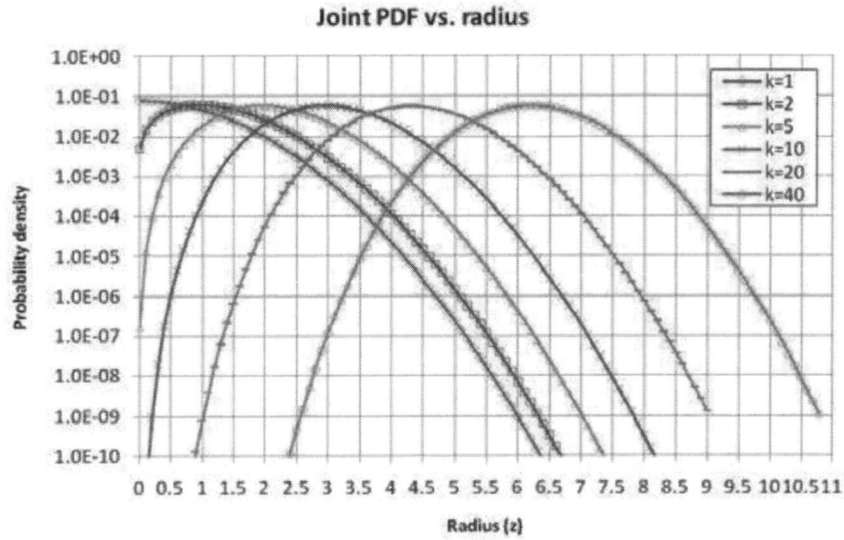

Figure 2. Joint PDF (9) vs. radius z for various k. Y scale is in log

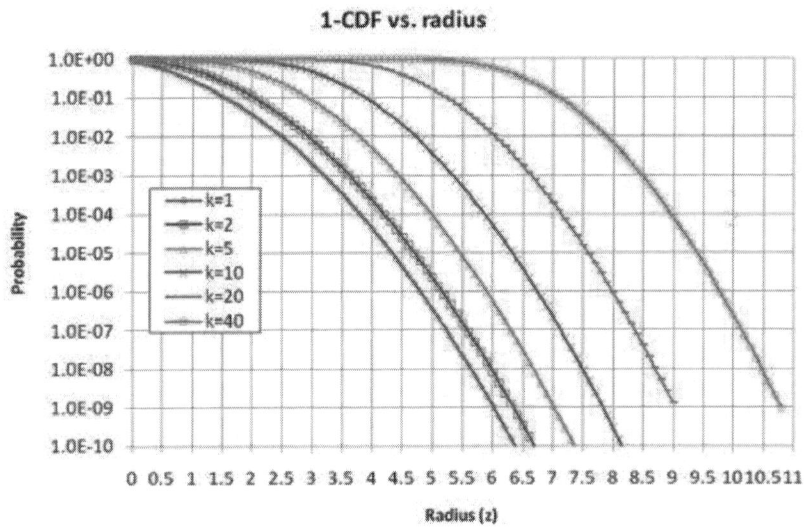

Figure 3. (1-CDF) vs. radius z for various k. Y scale is in log

For a probability of interest p_{tgt}, for example, finding the corresponding radius z_{tgt} requires solving (15), as graphically illustrated in Figure 3:

$$F\left(z_{tgt}\right)=\int_{z_{tgt}}^{+\infty}\frac{2^{1-k/2}}{\Gamma(k/2)}\cdot z^{k-1}\cdot e^{-\frac{z^2}{2}}\cdot dz=p_{tgt}$$

(15)

For example, z_{tgt} approximately equals 7.8 for a targeted probability of 10^{-9} and k=10 input variables (Figure 3).

This is the calculation used in XES to selectively refine the sampling in a range of radiuses corresponding to a range of probability of occurrence.

As in practice, we want to focus on given ranges of probability events far from the center of the parameter space (z=0), we will now design a sampling plan selectively enhancing the upper part of the CDF (15).

We underline here that, strictly speaking, low probability events can also happen at low radiuses for higher k values, as shown in Figure 2; these events are however very close to "nominal" conditions (i.e. all parameters at their average value for z=0) so are not really relevant to circuit analysis for which risks of functionality loss are expected for large deviations from typical conditions (high z values); note that it might not be the case in other particular contexts, in which low z values may contribute to lower and/or upper part of the CDF of the response and be the actual region of interest.

Using (15), lower and upper radiuses are calculated defining a region where the sampling refinement will be applied. Typically, the upper radius is taken at the lowest observable probability of occurrence in a given context (e.g. one over the number of cells in a memory array). In the following, the lower radius is varied depending on the objective of overall refinement of CDF or more selective refinement close to the upper radius, as illustrated in section 3.1. Outside of the refinement region, the sampling is relaxed. Optionally, the sampling can be extended beyond the upper radius (lowest probability of occurrence) in order to further refine the CDF near far quantiles.

As regards the evolution of the number of samples vs. radius, we chose a simple composition of linear growth and step functions. Other schemes are of course possible and would deserve a more in-depth investigation to determine an optimal *n*(z) function, particularly when k takes large values, since the area S_k (8) of k-spheres to be sampled grows rapidly with k ($\sim z^{k-1}$).

A typical sampling plan according to the above scheme is illustrated in Figure 4 in the case of 3 variables. The total number of samples is $\sim 2.10^4$ in this example. The lower boundary of sampling refinement region (region 2) is set to z_{low}=3.7 (corresponding to p_{low}=0.005), the upper boundary to z_{tgt}=6.8 ($p_{tgt}=10^{-9}$). In this case, 3% of the total number of samples were allocated to sampling beyond this limit, up to $p_{high}=p_{tgt}/10=10^{-10}$ (region 3). The dz parameter is set to 0.1.

An illustration of XES as compared to CMC sampling is shown in Figure 5 in the case of 2 random variables (N(0,1)). The isotropic sampling refinement for outer circles is the basic property of such a sampling strategy: without information on the response, no direction of space is favored. In this example, the coverage of parameter space with XES is almost reversed as compared to CMC, which is a typical sampling that XES can provide to outperform CMC.

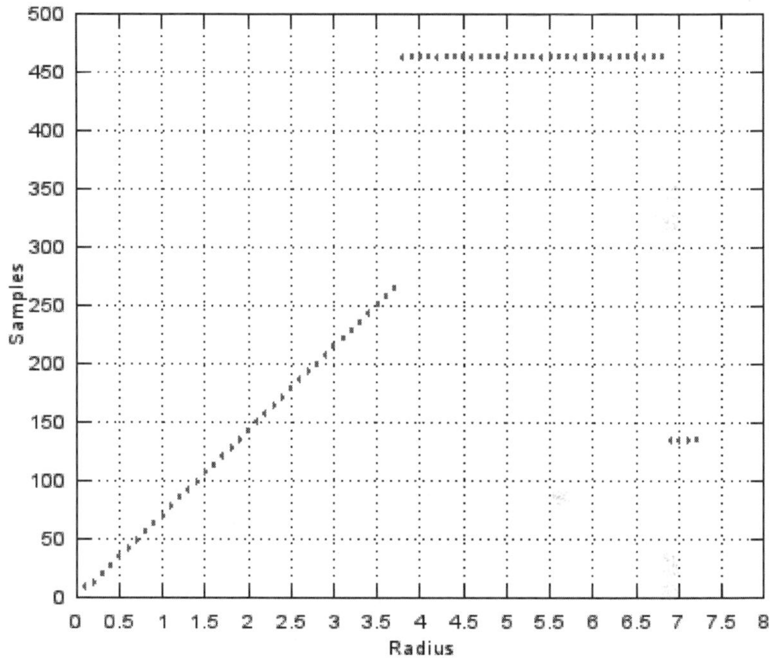

Figure 4. Sampling plan for 3 variables for a targeted probability of 10^{-9}. The lower boundary of sampling refinement corresponds to p_{low}=0.005, the upper boundary corresponds to $p_{tgt}=10^{-9}$; over-sampling beyond this radius aims at improving the CDF estimation in the vicinity of the targeted probability. The maximum radius corresponds to $p_{high}=10^{-10}$. dz=0.1

2.6. Actual Coordinates and Non-Normal Distributions

Up to this point, the random variables are still expressed in a standard normal scale. Therefore, for each variable, a transformation of coordinates from the standard normal distribution to their actual distribution has to be performed before acquiring (or simulating) the responses. For a normal distribution, this is simply achieved by restoring the proper standard deviation (σ) and mean (μ) along the k axes:

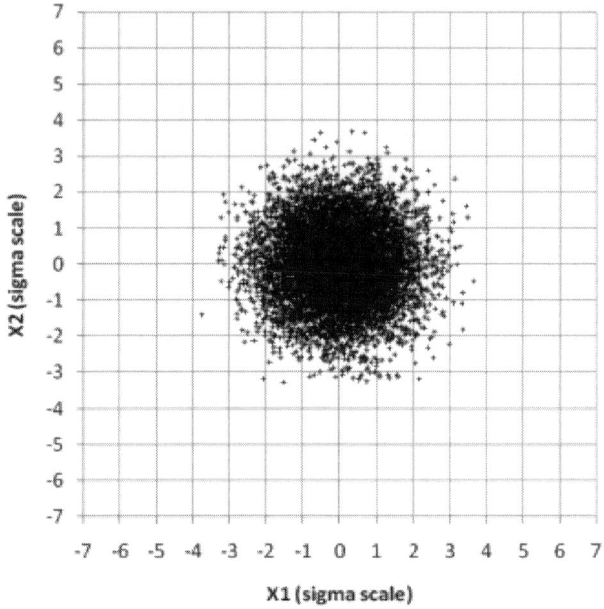

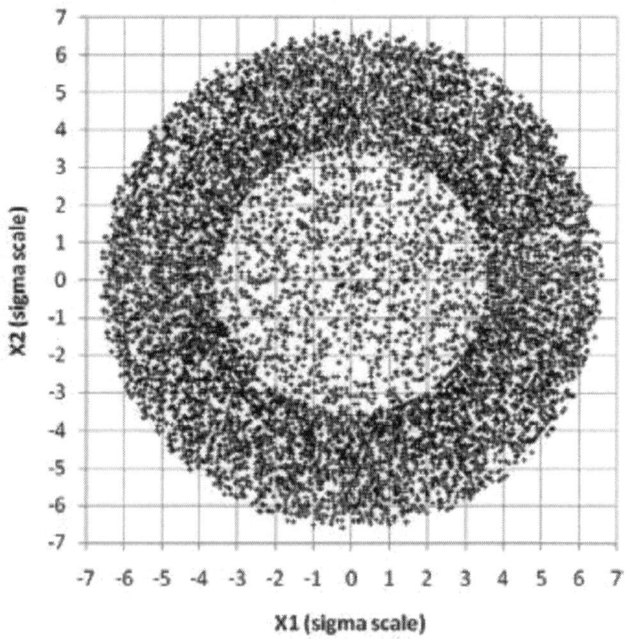

Figure 5. Illustration of sampling plans for 2 random normal variables N(0,1) for a/ CMC b/ XES strategies. Both cases have 10,000 samples. Targeted probability is $p_{tgt}=10^{-9}$, lower refinement boundary is $p_{low}=10^{-3}$. Scales are in sigma of the standard normal distribution

$$x_i = X_i \cdot \sigma_i + \mu_i \tag{16}$$

For non-normal distributions *D*, the *inverse transform method*[25] is applied to the relevant coordinates x_i:

$$x_i = F_D^{-1}(\Phi(X_i)) \tag{17}$$

Figure 6 illustrates how the sampling plan is modified by such a transformation for a normal distribution (along the horizontal axis) and a Weibull distribution (along the vertical axis) from the 2D sampling plan illustrated in Figure 5. The iso-probability contours are preserved by this transformation. As a result, in the example of Figure 6, it tends to concentrate more samples in the lower part of the space, due to the Weibull distribution shape (more visible in logarithmic scale). Within the XES strategy, this property ensures an even sampling, probability-wise, throughout the parameter space. However, the initial uniformity of samples on the k-spheres is not preserved by transformations (16) and (17), so the mean distance between samples on the new iso-probability surfaces is now varying (visible on the graph a/ of Figure 6), which tends to favor some portions of the space for a given probability of occurrence (no bias introduced, but coverage of space degraded). Searching a correction function to compensate for this modulation of density and thus further improve XES is certainly a topic of interest, but is however well beyond the frame of this work.

Finally, we stress here that the sampling itself is independent from the coordinate transformation (16) and (17). It implies that a same sampling plan can also potentially be used for various parameter distributions, which is of interest for example in a "what-if" analysis: studying the impact of changing the standard deviation of one of the variables does not require regenerating the sampling plan.

2.7. Confidence Interval, Error Calculation and Gains

To calculate a confidence interval (CI) on the estimated quantile ξ_p, we use the classical property of convergence in distribution of quantile error, as described in e.g.[26]:

$$\sqrt{N} \cdot \frac{f(\xi_p)}{\psi_p} \cdot \left(\tilde{\xi}_p - \xi_p\right) \Rightarrow \mathcal{N}(0,1) \tag{18}$$

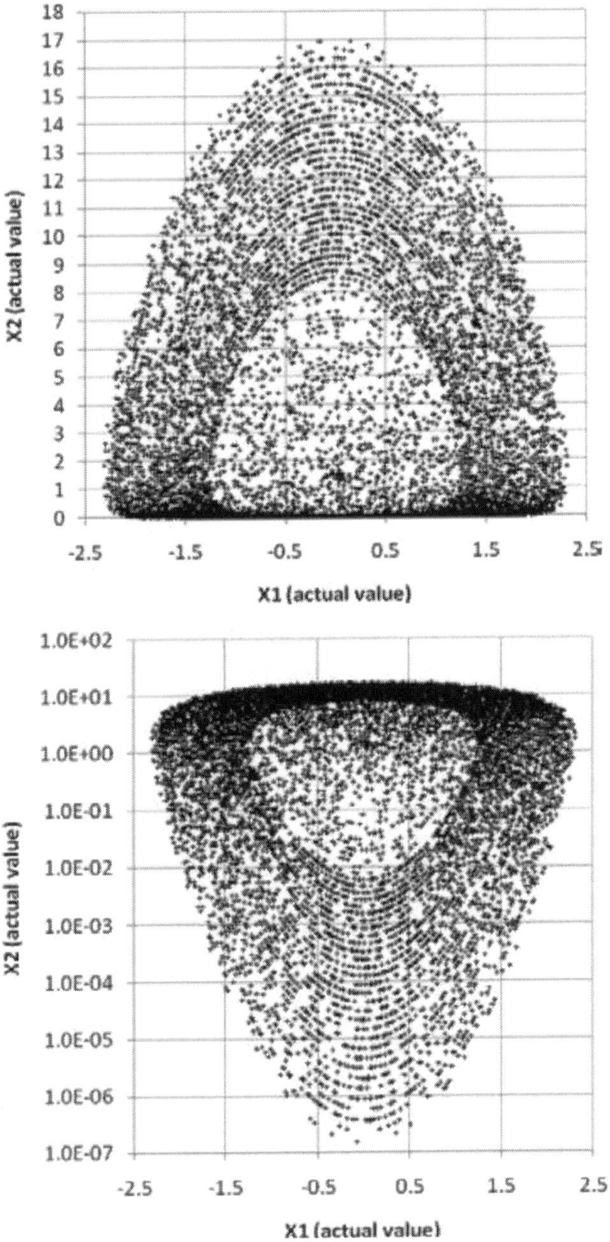

Figure 6. 2D sampling plan of Figure 5 after transformation of coordinates; X1: N(0,0.35^2); X2: Weibull (scale=2,shape=1.5). a/ linear Y-scale; b/ logarithmic Y-scale

where ψ_p is the standard deviation of the probability p (non-normalized to N) and $f(\xi_p)$ is the PDF of the response at the p-quantile, estimated in the following from the empirical CDF using a finite difference estimator as proposed in[26].

In the case of IS, the best estimator of the empirical variance ψ^2 depends on p and reads[26]:

$$\psi_p^2 = \left(\frac{1}{N} \sum_{i=1}^{N} \left(R\left(X_k^i\right) \leq \xi_p \right) \cdot \frac{W_i^2}{\overline{W}^2} \right) - p^2, p \leq 0.5$$
$$\psi_p^2 = \left(\frac{1}{N} \sum_{i=1}^{N} \left(R\left(X_k^i\right) > \xi_p \right) \cdot \frac{W_i^2}{\overline{W}^2} \right) - (1-p)^2, p > 0.5 \qquad (19)$$

For CMC, the asymptotic value of ψ^2 reads[26]:

$$\psi_p^2 = p \cdot (1-p) \qquad (20)$$

From (18), the standard deviation of the p-quantile estimator (ξ_p) reads:

$$\sigma_{\xi_p} = \frac{1}{\sqrt{N}} \cdot \frac{\psi_p}{f(\xi_p)} \qquad (21)$$

where ψ_p is calculated from (19) for XES or (20) for CMC.

Then, to calculate a 95% CI, we used the classical approximation:

$$CI_{95\%} = \tilde{\xi}_p \pm 1.96 \cdot \sigma_{\xi_p} \qquad (22)$$

In order to estimate the accuracy of estimated probabilities, we also calculate the relative error ε_p on a given probability p[1]:

$$\varepsilon_p = \frac{\sigma_p}{p} = \frac{1}{N} \cdot \frac{\psi_p}{p} \qquad (23)$$

To further check the validity of the above confidence interval estimate (22) in the case of XES, we experimentally compared it to a CI calculated by re-sampling through a bootstrap (BS) technique and using Studentized statistics [27][28]. As we will illustrate in the next section, a good agreement is observed between the CIs calculated by the two methods.

Finally, we define two gains (Ga, Gs) to measure the performance of XES over CMC:

$$G_a = \frac{\varepsilon_p(CMC)}{\varepsilon_p(XES)}$$
$$G_s = \frac{N_{CMC}(\varepsilon_p \le 0.15)}{N_{XES}(\varepsilon_p \le 0.15)} \tag{24}$$

G_a being the gain in accuracy at a given sampling N, and G_s the gain in sampling at a given relative error ε_p(15% in this case). We can notice that the gain Ga is the ratio of relative errors, which is actually equivalent to a classical metric measuring the gain of an IS over CMC at a given p-quantile for a constant number of samples N: Ga~σ_{MC}/σ_{IS}[4].

3. EMPIRICAL STUDY

We implemented the above formulas and sampling scheme in a commercial numerical computing platform in order to evaluate the accuracy and efficiency of the XES method, as we will now illustrate. The examples shown hereafter are built around a series of simple electrical test circuits made of resistors connected in series or in parallel and biased under a constant current Ic (Figure 7). The studied response is a voltage drop (Vd) across the circuit, and the random variables are the resistances. These simple circuits could represent, for example, a memory cell that would be repeated over a memory array and would experience statistical variations of its parameters (here: the resistance value).

Since the equivalent resistance of such circuits is calculable, the response has a known analytical formula (Figure 7). In the following examples, this formula is used to calculate the response Vd for each sample and hence reconstruct the CDF; in an actual implementation, a circuit solver would compute the responses.

Notes: probabilities in the CDF graphs below are expressed in sigma of the standard normal distribution, except otherwise specified. On such a scale, purely normal distributions show up as a straight line. The parameter dz in (12) is set to a constant value of 0.1 in all the examples below.

3.1. Linear Response

We will first illustrate XES in the case of a Type 1 circuit (Figure 7 a/) for which the response is linear with respect to the resistances, and for 3 resistors: R1, R2, R3, following normal (section 3.1.1.) or non-normal (section 3.1.2.) distributions.

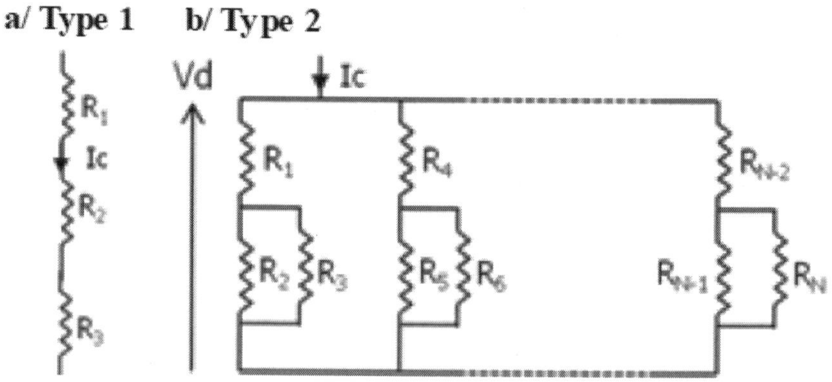

Figure 7. Test circuit schematics with resistors in a/ series (s) (Type 1) or b/ series (s) and parallel (//) (Type 2). The equivalent resistance Req can be calculated in all cases using the classical formulas Req(Rx.s.Ry)=Rx+Ry Req(Rx//Ry)=Rx.Ry/(Rx+Ry). The studied response is the voltage drop Vd across the group of resistors under a constant current Ic (Ic=1mA in all the following examples): Vd= Req Ic

3.1.1. Normal Distributions

The resistors are chosen as follows: R1:N($50,0.5^2$); R2: N($100,10^2$); R3:N($120,1.5^2$) – all values in ohm. The number of samples was set to 13,896 for the various plots below.

Figure 8 shows the resulting XES CDF for Vd with the p_{low} parameter set to 10^{-6} and for a targeted probability of 10^{-9} (assuming the studied population is of 1 billion individuals). Since in this case, the theoretical CDF can also be calculated from the composition of variance[18], we also plotted it in the same graph, in order to check that the two traces were superimposed.

The 95% CI was also computed and is shown in Figure 9 together with the CDF for the lower (a/) and upper (b/) part of the CDF. The CI calculated with (22) and from a bootstrap re-sampling gives consistent results. As expected, thanks to the XES strategy, the CI is narrow in the vicinity of the targeted probability, both at +6σ and at -6σ.

The relative error on the estimated quantiles (δξ) can also be calculated in this specific case, since the theoretical CDF and hence the true ξ_p are known:

$$\delta_\xi = \frac{\left|\tilde{\xi}_p - \xi_p\right|}{\xi_p} \qquad (25)$$

With the above sampling plan, the measured δξ at ±6σ is well below 1% (typically around 0.1-0.2%).

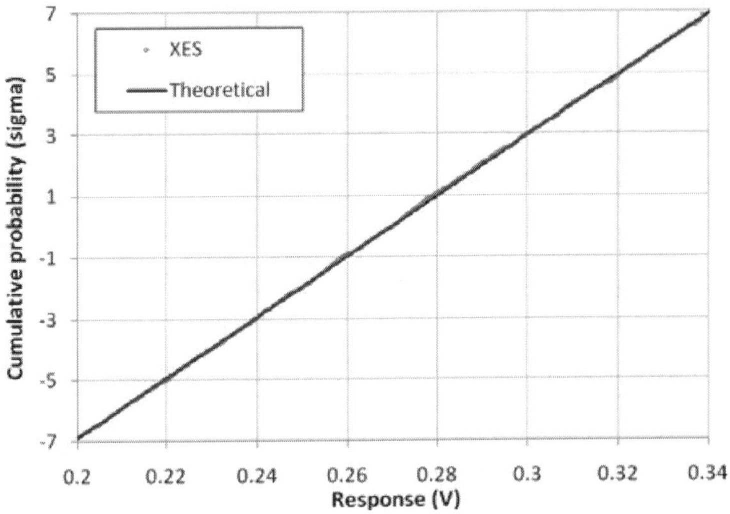

Figure 8. CDF (sigma scale) of voltage drop (Vd) for 3 resistors / Type 1 circuit. The p_{tgt}parameter is set to 10^{-9} and p_{low} to 10^{-6}. XES CDF is superimposed to the theoretical straight line

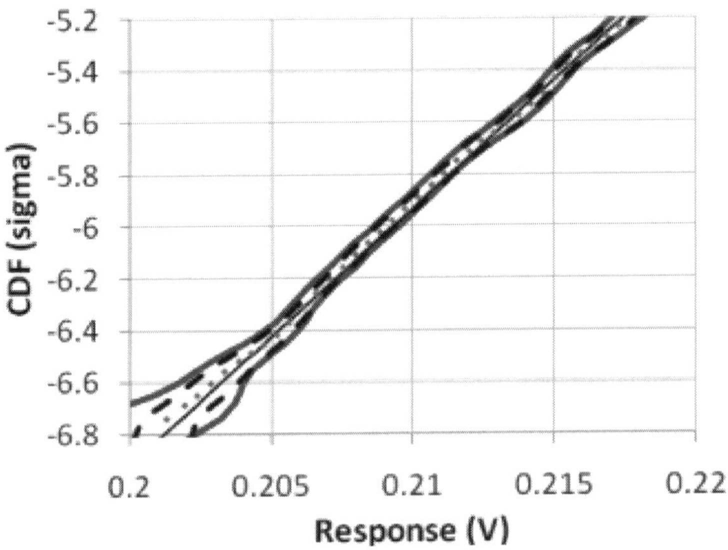

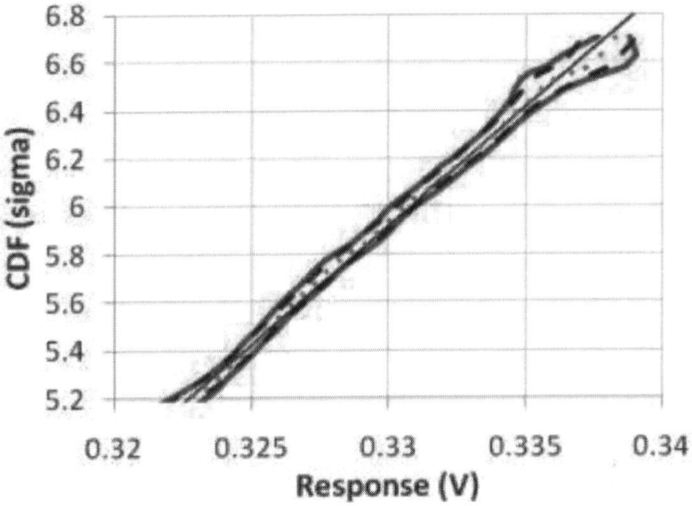

Figure 9. Detail of a/ lower and b/ upper part of the XES CDF of Vd from Figure 8 as compared to the theoretical straight line; 95%CI calculated from (22) or from BS re-sampling (1000 runs)

The effect of modulating the refinement region (Figure 10) through the p_{low} parameter at a constant number of samples on the error ε_p is illustrated in Figure 11. The peak relative error observed in the 4-5σ range tends to decrease and to shift to lower σ for increasing p_{low}, while the relative error degrades at the targeted probability of ±6σ. In this example, the criterion of $\varepsilon_p < 15\%$ at ±6σ is reached for plow=10^{-6}.

Another property of XES is evidenced in Figure 11: because of the symmetry of the sampling on k-spheres, both ends of the CDF are refined by over-sampling rare events corresponding to large z radiuses, leading to an accurate estimation of p_{tgt} (-6σ) and 1-p_{tgt} (+6σ) together. This is a key feature to analyze responses showing a two-sided operating range (i.e. both low and high values leading to circuit failure). As already mentioned before, we exclude from this study the particular cases of responses exhibiting far quantiles at low z-values (specific non-monotonic functions), for which the above sampling refinement would not be effective on the lower (or upper) side of the CDF.

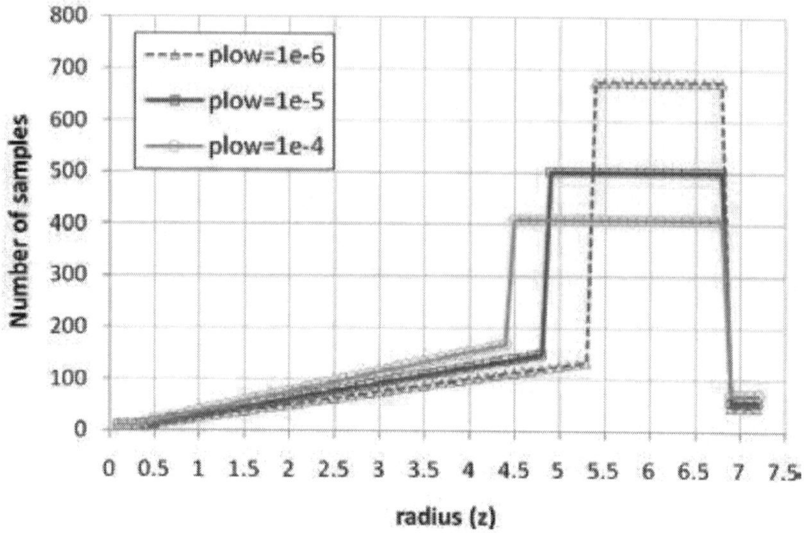

Figure 10. Sampling plan for various plow, at a constant N=13,896

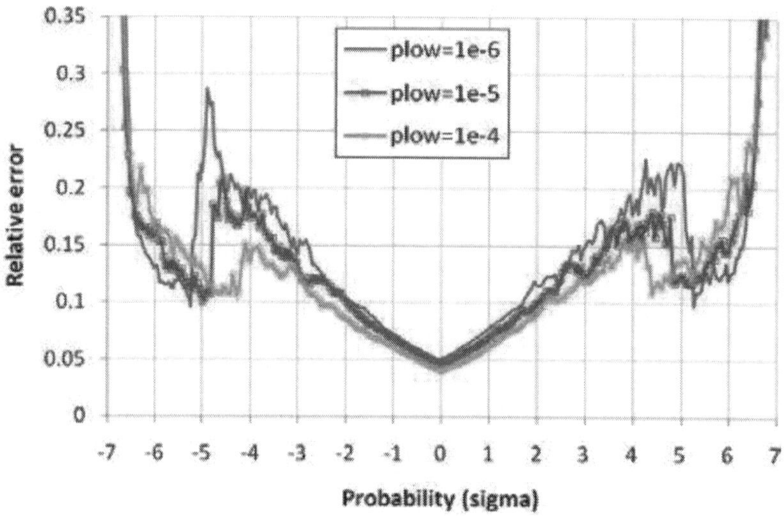

Figure 11. Relative error εp vs. probability for various plow values at a constant N for CDF of Figure 8. The variations of plow modulate the refinement sampling range and the accuracy in this region

3.1.2. Non-Normal Distributions

In this example, the circuit is identical to the previous (3 resistors in series) and resistances are simply changed to 3-parameter Weibull (W) distributions instead of normal, as follows: W_{R1}(scale=1, shape=1.5, location=50), W_{R2}(scale=

1, shape=30, location= 100), W_{R3}(scale=1, shape=4.5, location=120). The values were chosen to yield various PDF asymmetries, as illustrated in Figure 12. The XES parameters are: N=13,872 samples, $p_{tgt}=10^{-9}$, $p_{low}=10^{-4}$. The resulting XES CDF is shown in Figure 13 and is superimposed to the CMC trace. The confidence interval in the lower and upper part of the CDF are shown in Figure 14, further illustrating the good accuracy obtained at extreme quantiles even for non normal distributions.

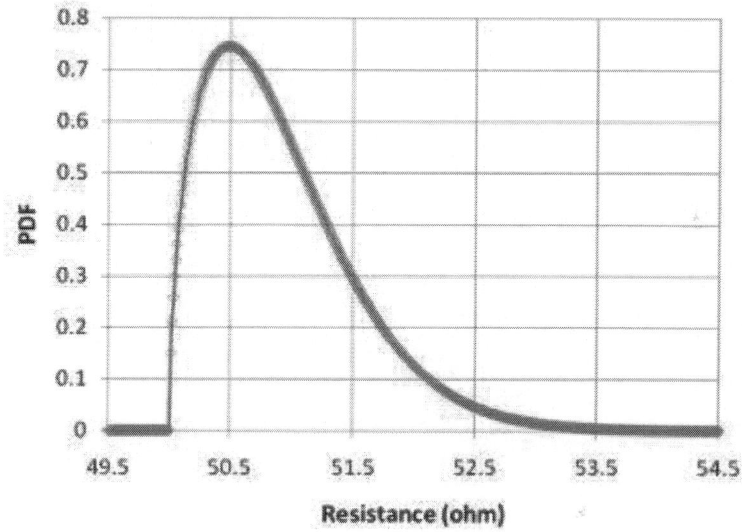

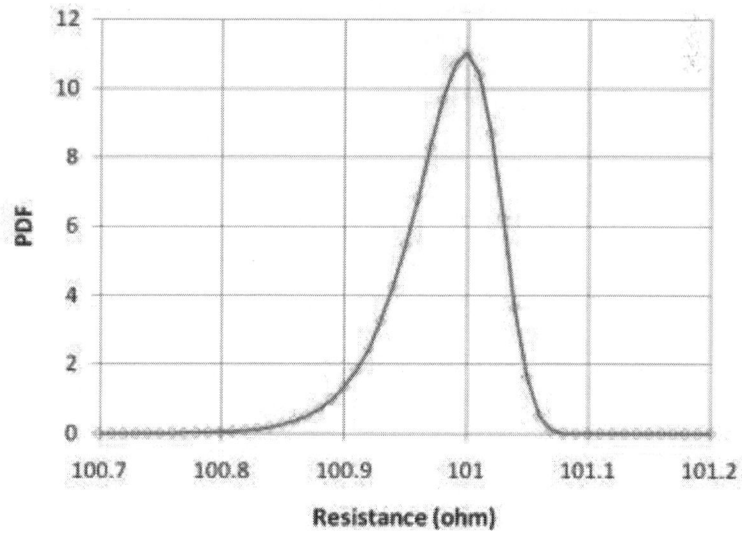

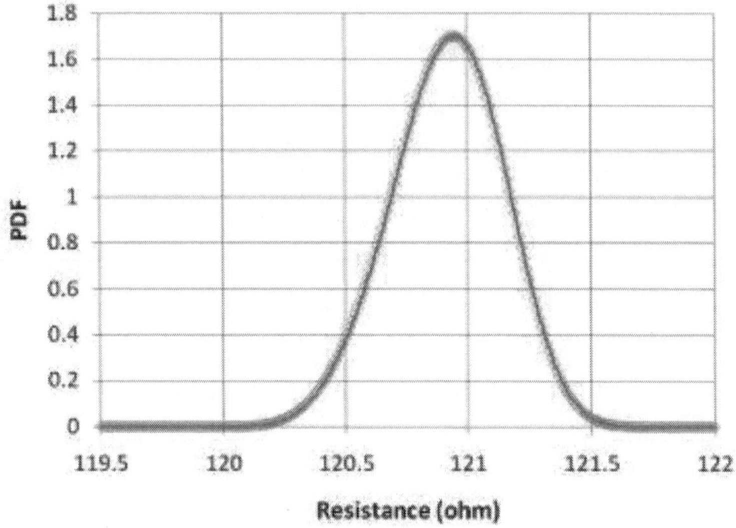

Figure 12. PDF of the resistances R1, R2 and R3 (Weibull distri butions)

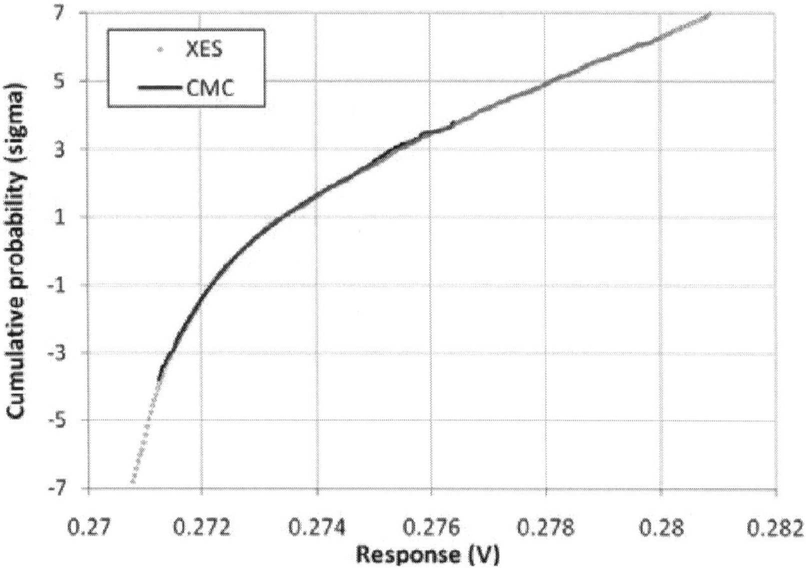

Figure 13. XES and CMC CDF of the response Vd of 3 resistances following 3-parameter Weibull distributions, Type 1 circuit

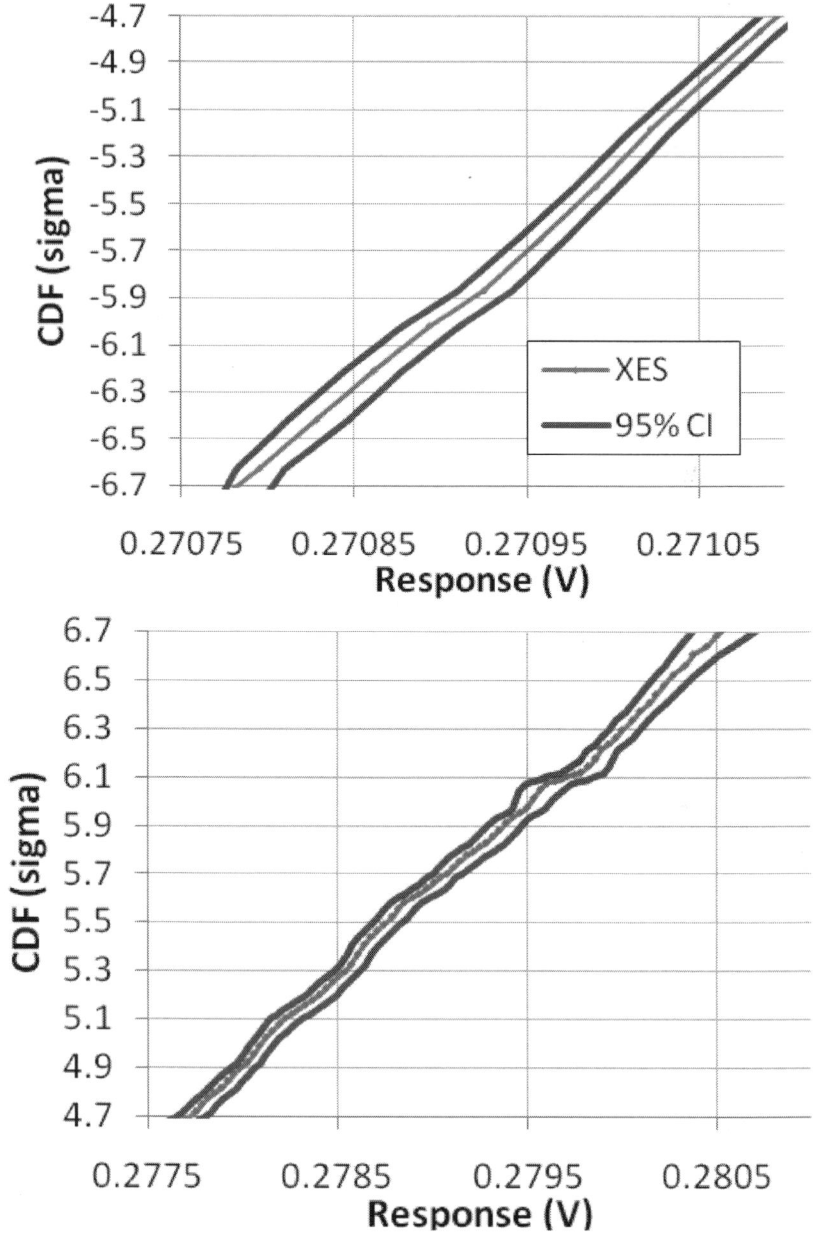

Figure 14. XES CDF of Vd and 95% Confidence Interval from (22) for a/ lower and b/ upper part of the CDF

3.2. Non Linear Response

Considering now a Type 2 circuit and three normally distributed resistances, the resulting response Vd is non-linear, due to the resistors in parallel (Figure 7). In the following example, the resistances are chosen identical to the ones of section 3.1.1. The sampling plan is also similar, with $p_{tgt}=10^{-9}$, $p_{low}=10^{-5}$ and N=13,880 samples. The CDF of the response are shown in Figure 15 together with the 95% CI calculated from (22) or with 1000 BS runs. Tight CIs are obtained in the targeted probability range and the two CIs are consistent.

Finally, the relative error (ε_p) curve shown in Figure 16 clearly demonstrates the effect of the sampling refinement on the accuracy and illustrates that ε_p~15% at $p_{tgt}=10^{-9}$ is reached for these sampling conditions, hence proving the effectiveness of XES also for non linear responses.

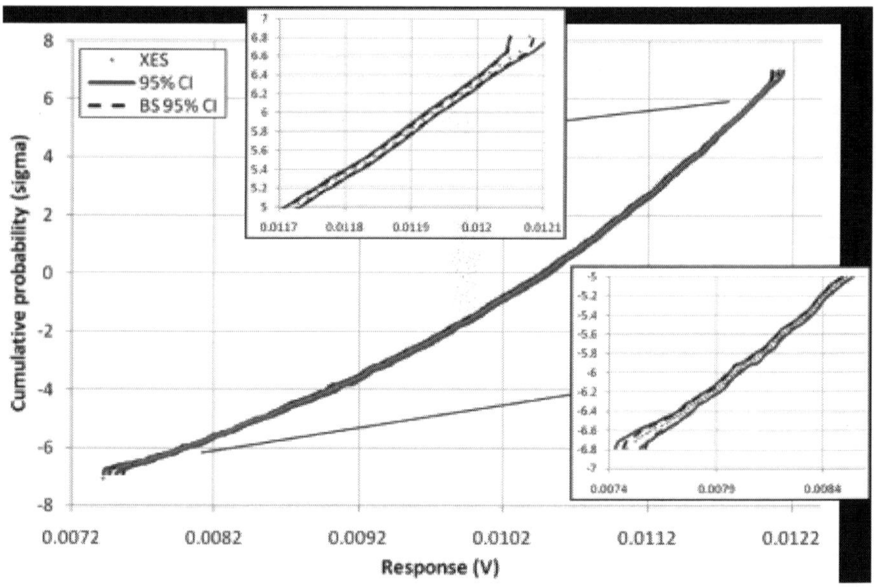

Figure 15. XES CDF of Vd and 95% confidence interval from (22) and from BS using 1000 runs (insets: detail of the a/ upper and b/ lower part of the CDF). Type 2 circuit with 3 resistors identical to section 3.1.1

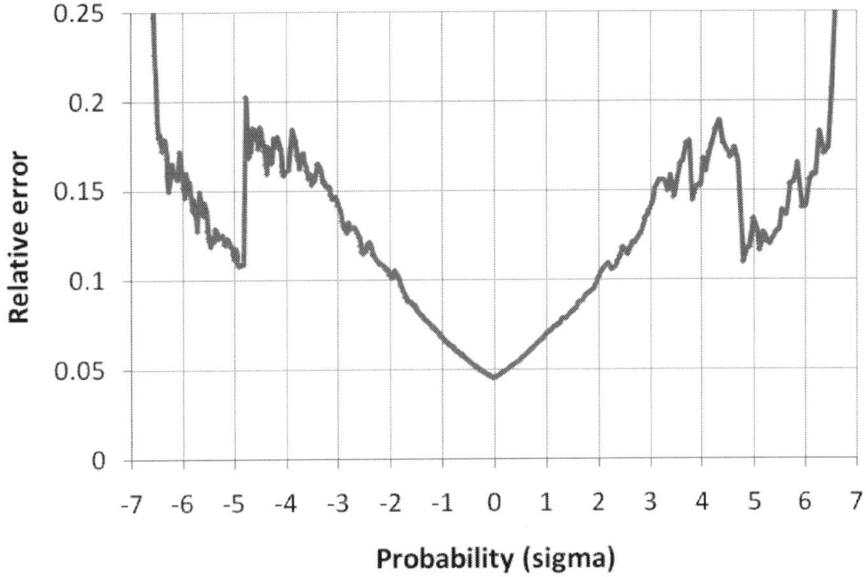

Figure 16. Relative error ε_p vs. cumulative probability (sigma scale)

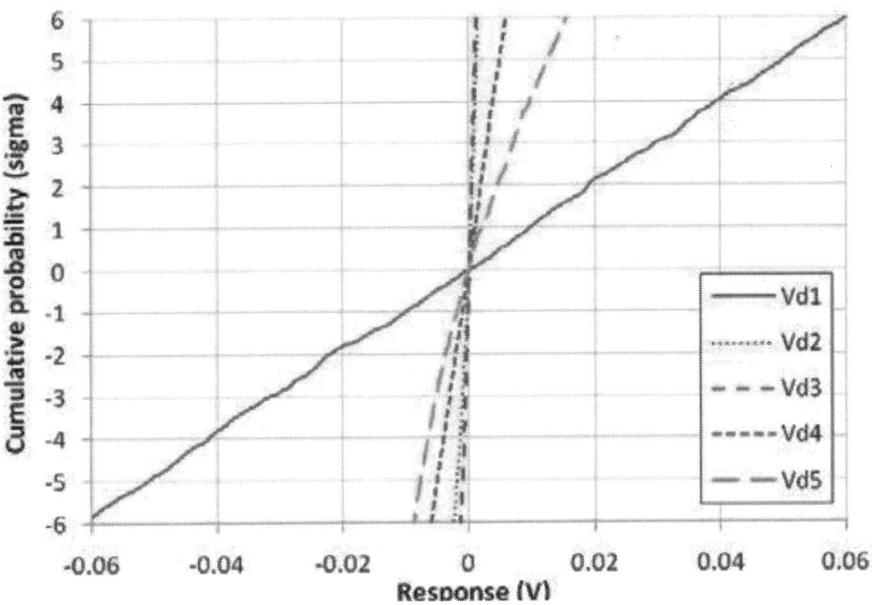

Figure 17. CDF of various responses, single sampling plan, p_{low}, $=10^{-4}$, N=13,989. The responses have been normalized (shifted to their median) in order to represent the various responses on the same x-scale

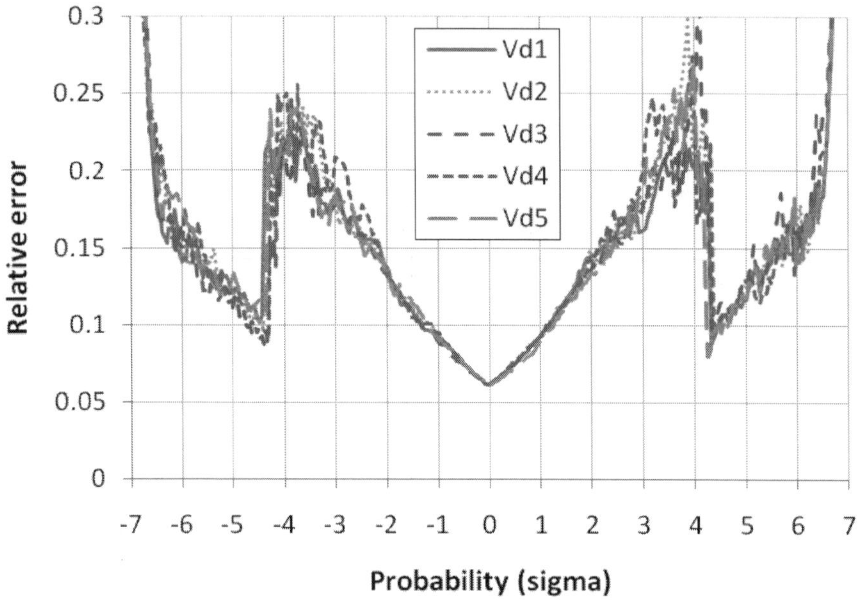

Figure 18. Relative error ε_p for the various responses Vd1 to Vd5, single sampling plan, p_{low}, $=10^{-4}$, $p_{tgt}=10^{-9}$, N=13,989

3.3. Multiple Responses, Single Sampling Plan

In cases where the response has a known analytical formula as in the above examples, IS approaches such as*exponential twisting*[29] or *mixture IS*[8], for example, would certainly be more efficient, but restricted to one response at a time, as the sampling is adjusted to the response shape (either by knowing the response beforehand, or through a feedback of the response on the sampling while running). On the contrary, XES is built to deal with multiple arbitrary responses with a single set of samples, as we will now further illustrate on an example.

Let's consider 5 different responses of 3 normally distributed variables, built as follows: Vd1 is a linear response (Type 1 circuit) identical to section 3.1.1. Vd2 is identical to section 3.2 (Type 2 circuit), Vd3 and Vd4 are also from a Type 2 circuit but with a cyclic permutation of resistances as compared to Vd2. Vd5 is a purely fake response, built from the sum of square of resistances multiplied by a constant (10^{-6}), in order to illustrate the case of a response that is non linear and non-monotonic with respect to all its variables.

As shown in Figure 17, despite the collection of responses yielding a large variety of CDF shapes, XES allows calculating all of them from a single set of samples. Moreover, in the sampling refinement region ($p_{low}=10^{-4}$; $p_{tgt}=10^{-9}$), the relative error ε_p is low and is similar for the various responses, as illustrated in

Figure 18, proving the effectiveness and robustness of the method to arbitrary response shapes.

3.4. Larger Number of Variables (k)

As mentioned in section 2.5, the area of a k-sphere is growing rapidly with k and z, hence the number of samples required by XES is expected to grow accordingly for far quantiles.

We have empirically studied the evolution of the minimal sampling to reach a given relative error (ε_p<15%) at a target quantile (for both p_{tgt} and its counterpart 1-p_{tgt}), using a simple response (Type 1 circuit) and a growing number of normally distributed variables (k). As graphically illustrated in Figure 19 for two target quantiles ($p_{tgt}=10^{-9}$ and 10^{-10}), the growth of the number of samples vs. k is indeed fast, especially in the initial range of 3-12 variables.

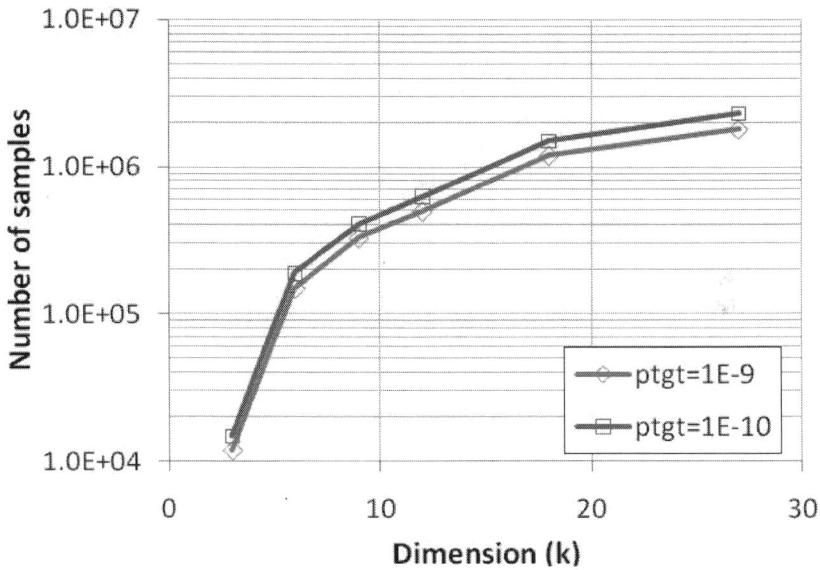

Figure 19. Required number of samples to reach ε_p<15%@p_{tgt} vs. number of variables (k) for two target quantiles, 10^{-9} and 10^{-10}. Response is Vd of a Type 1 Circuit; all variables are normally distributed

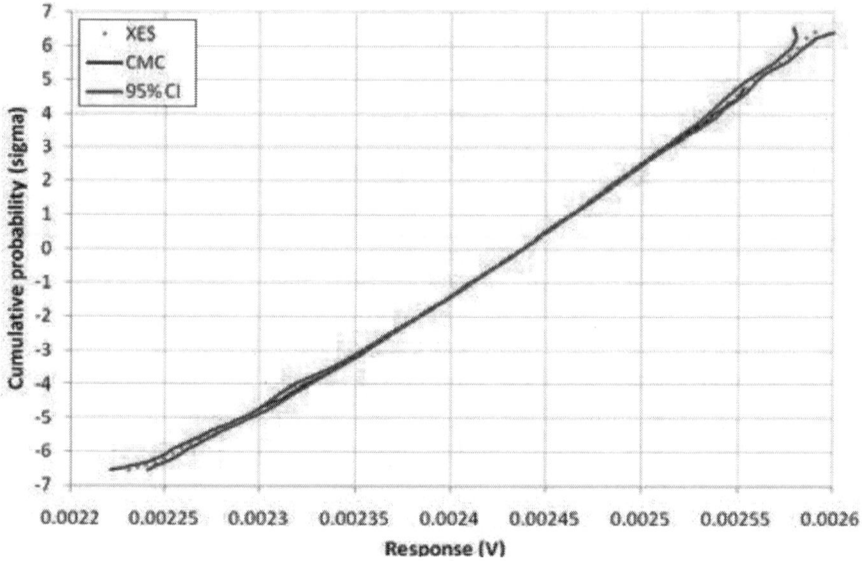

Figure 20. XES and CMC CDF of a response (Vd) of 18 variables; Type 2 circuit with resistance mean ∈[50 100 120 60 130 90 70 110 150 140 80 110 105 70 90 85 108 160] and standard deviation ∈[0.5 10 1.5 0.7 10 3.5 0.5 3 5.5 5.5 3 2.8 7.8 1.8 2.2 5 9 3]. 95% CI calculated from (22). $p_{tgt}= 10^{-9}$ $p_{low}=10^{-3}$, $N=1.4.10^{6}$ samples

We also verified that a CDF of a non linear response (Type 2 circuit) could also be computed with a reasonable accuracy in similar sampling conditions, as illustrated in Figure 20 for 18 variables. In this example, the sampling is optimized to keep a tight CI over the full CDF rather than yielding a given accuracy for a given quantile.

3.5. Benchmark to CMC

Based on the previous results, we will now further illustrate the efficiency of XES as k grows by comparing its performances to CMC in term of gain in sampling (Gs) and in accuracy (Ga) as defined in (24).

The Tables 1 and 2 below show the gains obtained for $p_{tgt}=10^{-9}$ and 10^{-10}, respectively, and for various k (up to k=27), considering a response Vd from a Type 1 circuit. The same data are graphically illustrated in Figure 21 and 22, showing that both gains stay quite strong even for 27 variables and further grow as the target quantile is decreasing to $p_{tgt}=10^{-10}$. It also shows that the gain degradation tends to attenuate as k grows, suggesting that the efficiency of XES over CMC should stay significant for an even larger number of variables.

We finally underline that the error on the estimated quantile $\delta\xi$ at p_{tgt} was below 1% for all the XES results of Tables 1 and 2, proving that the method does not introduce significant biases in the far quantile estimates.

Table 1. Expected gains as compared to CMC – various p_{low}, $p_{tgt}=10^{-9}$

k	Sampling (N)	Gain vs. CMC at $p_{tgt}=10^{-9}$	
		Gs: sampling gain at 15% relative error on p_{tgt}	Ga: accuracy gain at given sampling
3	1.2E+04	4.2E+06	1924.5
6	1.5E+05	3.3E+05	544.3
9	3.30E+05	1.5E+05	367.0
12	5.0E+05	1.0E+05	298.1
18	1.2E+06	4.2E+04	210.8
27	1.8E+06	2.8E+04	157.1

Table 2. Expected gains as compared to CMC – various p_{low}, $p_{tgt}=10^{-10}$

k	Sampling (N)	Gain vs. CMC at $p_{tgt}=10^{-10}$	
		Gs: sampling gain at 15% relative error on p_{tgt}	Ga: accuracy gain at given sampling
3	1.5E+04	3.3E+07	5443.3
6	1.9E+05	2.6E+06	1529.4
9	4.10E+05	1.2E+06	1041.2
12	6.3E+05	7.9E+05	839.9
18	1.5E+06	3.3E+05	544.3
27	2.30E+06	2.2E+05	439.6

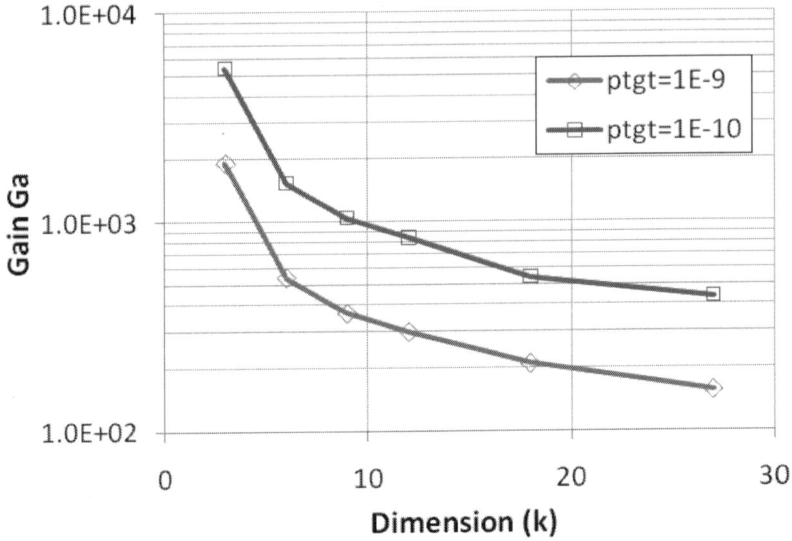

Figure 21. Gain in accuracy of XES over CMC vs. k for two target quantiles $p_{tgt}=10^{-9}$ and 10^{-10} and at a given sampling (see Tables 1 and 2). Response is Vd of a Type 1 Circuit; all variables are normally distributed

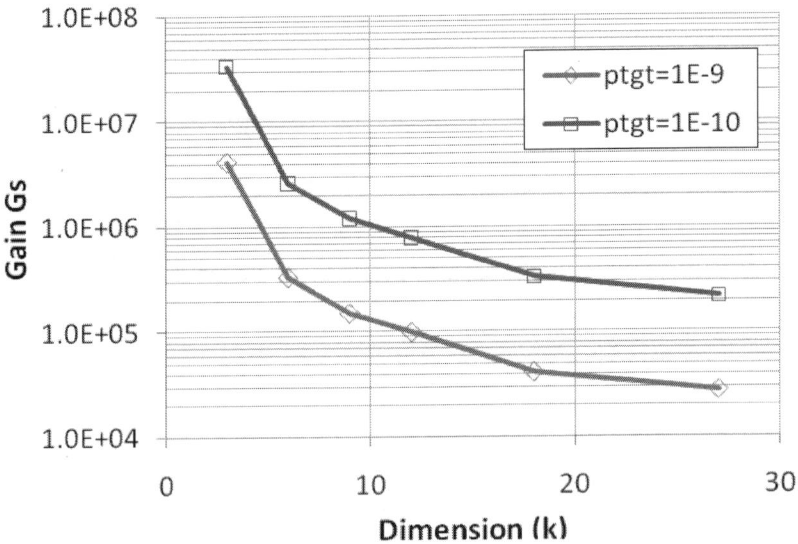

Figure 22. Gain in sampling of XES over CMC vs. k for two target quantiles $p_{tgt}=10^{-9}$ and 10^{-10} and for $\varepsilon_p<15\%$. Response is Vd of a Type 1 Circuit; all variables are normally distributed

4. CONCLUSIONS

We propose a multivariate sampling method, called XES (eXtreme Event Sampling), as an alternative to CMC and classical IS to compute far quantiles of arbitrary responses. The proposed strategy is similar to an IS scheme, but unlike classical IS techniques, does not require to know the response in advance or to adjust the sampling plan to it. Instead, events of low probability of occurrence are over-sampled in order to enhance far quantiles of responses.

As a result, CDF of arbitrary responses can be reconstructed with a good accuracy at extreme quantiles and much more efficiently than CMC, despite the fact the method is generic as compared to an IS approach.

The method applies to normally or non-normally distributed parameters and can deal with a large number of variables (up to 27 tested).

Thanks to its flexibility, we believe XES is a valuable alternative to classical IS strategies when multiple arbitrary responses are to be studied in a single-pass simulation run (i.e. without adapting the sampling from the responses). Thus, beyond the case of circuit design of interest to us, the method could apply to other domains showing similar constraints.

Future developments of the method should include in-depth investigation of the optimal sampling strategy (n(z) function) vs. k, further benchmark including to other IS strategies and finally, implementation of XES in a circuit solver to study real cases.

ACKNOWLEDGEMENTS

The authors would like to acknowledge Alessandro Calderoni and Andrea Marmiroli (Micron Technology Italy) for their kind support and for valuable discussions and suggestions during the development of the XES method.

REFERENCES

1. M. Pagano, W. Sandmann, "Efficient Rare Event Simulation - A Tutorial on Importance Sampling", in Proceedings of the 3rd International Conference on Performance Modelling and Evaluation of Heterogeneous Networks (HetNets), pp.T03/1-38, 2005.
2. R. Srinivasan, Importance sampling - Applications in communications and detection, Springer-Verlag, Germany, 2002.
3. Tim C. Hesterberg, "Advances in Importance Sampling", PhD dissertation, Stanford University (CA), USA, 1988 (revised 2003).

4. P.J. Smith, M. Shafi, H. Gao, "Quick simulation: a review of importance sampling techniques in communications systems", IEEE Journal on Selected Areas in Communications, vol.15 no.4, pp.597-613, 1997.

5. P. Glasserman, Y. Wang, "Counter Examples in Importance Sampling for Large Deviations Probabilities", Institute of Mathematical Statistics, The Annals of Applied Probability, vol.7, no.3, pp.731-746, 1997.

6. J. Yeung, H. Mahmoodi, "Robust sense amplifier design under random dopant fluctuations in nano-scale CMOS technologies", in Proceedings of IEEE International SOC Conference, p.261, 2006.

7. S. Akiyama, T. Sekiguchi, K. Kajigaya, S. Hanzawa, R. Takemura, T. Kawahara, "Concordant memory design using statistical integration for the billions-transistor era", in International Solid-State Circuits Conference Digest, p.466, 2005.

8. Rouwaida Kanj, Rajiv Joshi, Sani Nassif, "Mixture importance sampling and its application to the analysis of SRAM designs in the presence of rare failure events", in Proceedings of IEEE Design Automation Conference, pp.69-72, 2006.

9. H. Nho, S.-S. Yoon, S. Wong, S.-O. Jung, "Statistical simulation methodology for sub-100nm memory design", IET, Electronics Letters, vol.43, no.16, pp.869-870, 2007.

10. D. Reid, C. Millar, G. Roy, S. Roy, A. Asenov, "Analysis of threshold voltage distribution due to random dopants: a 100 000-sample 3-D simulation study", IEEE Transaction on Electron Devices, vol.56, no.10, pp.2255-2263, 2009.

11. P. Magnone, A. Mercha, V. Subramanian, P. Parvais, N. Collaert, M. Dehan, S. Decoutere, G. Groeseneken, J. Benson, T. Merelle, R. J. P. Lander, F. Crupi, C. Pace, "Matching performance of FinFET devices with Fin widths down to 10 nm", IEEE Electron Device Letters, vol.30, no.12, p.1374, 2009.

12. S. Toriyama, D. Hagishima, K. Matsuzawa, N. Sano," Device simulation of random dopant effects in ultra-small MOSFETs based on advanced physical models", in Proceedings of SISPAD conference, p.111, 2006.

13. A. Asenov, "3D statistical simulation of intrinsic fluctuations in decanano MOSFETS induced by discrete dopants, oxide thickness fluctuations and LER", in Proceedings of Simulation of Semiconductor Processes and Devices, pp.162-169,2001; online:http://in4.iue.tuwien.ac.at/pdfs/sispad2001/pdfs/AsenovA_36.pdf.

14. Overall Roadmap Technology Characteristics Tables, International Technology Roadmap for Semiconductors (ITRS), 2011 edition;

15. Koren, C. M. Krishna, Fault-Tolerant Systems, Elsevier, USA, 2007.

16. C. Caillat, E. Carman, "A fast and accurate statistical sampling technique to predict the impact of device fluctuations on the functionality of large memory arrays", in Proceeding of the 2011 Micron Technical Seminar (internal Micron publication), pp.36-38, 2012.

17. C. Caillat, E. Carman, "A Multivariate Statistical Sampling Technique (MuSST) to Assess the Robustness of Very Large Memory Arrays to Local Fluctuations", unpublished, 2012.

18. D.P. Bertsekas, J.N. Tsitsiklis, Introduction to Probability, 2nd ed., Athena Scientific, USA, 2008.

19. L. E. Blumenson, "A Derivation of n-Dimensional Spherical Coordinates", Mathematical Association of America, The American Mathematical Monthly, vol.67 no.1, pp.63-66, 1960.

20. D.M.Y. Sommerville, An Introduction to the Geometry of n Dimensions, Dover Publications, USA, 1958.

21. Harald Niederreiter, Random Number Generation and Quasi-Monte Carlo Methods, Society for Industrial and Applied Mathematics, USA, 1992.

22. Peter W. Glynn, "Importance Sampling for Monte Carlo Estimation of Quantiles", in Proceedings of the Second International Workshop on Mathematical Methods in Stochastic Simulation and Experimental Design, pp.180-185, 1996.

23. G. Marsaglia, "Choosing a point from the surface of a sphere", Institute of Mathematical Statistics, The Annals of Mathematical Statistics, vol.43. no.2, p.645, 1972.

24. Aruna Sivakumar, Chandra R. Bhat, Giray Ökten, "Simulation Estimation of Mixed Discrete Choice Models Using Randomized Quasi-Monte Carlo Sequences: A Comparison of Alternative Sequences, Scrambling Methods, and Uniform-to-Normal Variate Transformation Techniques", Transportation Research Board, Transportation Research Record, vol.1921, pp.112-122, 2005.

25. George Casella, Roger L. Berger, Statistical Inference, 2nd ed., Duxbury Press, USA, 2001.

26. Fang Chu, Marvin K. Nakayama, "Confidence Intervals for Quantiles and Value-at-Risk when Applying Importance Sampling", in

Proceedings of the 2010 Winter Simulation Conference, pp.2751-2761, 2010.

27. A. C. Davison, D. V. Hinkley, Bootstrap Methods and their Application, Cambridge University Press, USA, 1997; Shortcourse (Padova University, Italy, 2006) based on the book available at the link:http://www.stat.unipd.it/uploads/File/archivio/20060920141934_20060731132050001_Materiale_Didattico_Davison.pdf.

28. Bradley Efron, The Jackknife, the Bootstrap, and Other Resampling Plans, Society for Industrial and Applied Mathematics, USA, 1987. Original technical report from the Department of Statistics, Stanford University (CA), USA, 1980 is available online:http://statistics.stanford.edu/~ckirby/techreports/BIO/BIO%2063.pdf.

29. Paul Glasserman, Sandeep Juneja, "Uniformly Efficient Importance Sampling for the Tail Distribution of Sums of Random Variables", Journal of Mathematics of Operations Research archive, vol.33, no.1, pp.36-50, 2008.

Chapter 10

On Posterior Analysis of Mixture of Two Components of Gumbel Type II Distribution

Navid Feroze [1], **Muhammad Aslam** [2]

[1]Department of Mathematics and Statistics, Allama Iqbal Open University, Islamabad, Pakistan

[2]Department of Statistics, Quaid-i-Azam University, Islamabad, Pakistan

ABSTRACT

This paper describes the Bayesian analysis of the parameters of mixture of two components of Gumbel type II distribution. A heterogeneous population has been modeled by means of two components mixture of the Gumbel type II distribution under type I censored data. The Bayes estimators of the said parameters have been derived under the assumption of non-informative priors on the basis of different loss functions. A censored mixture data is simulated by probabilistic mixing for the computational purpose. The comparisons among the estimators have been made in terms of corresponding posterior risks. The posterior predictive distributions and intervals have been derived and evaluated under each prior.

Keywords: Bayes Estimators, Posterior Risks, Mixture Models, Loss Functions

1. INTRODUCTION

The Gumbel type II distribution is used to model the extreme events like extreme earthquake, rainfalls, temperature, floods etc. It has another side of applications which deals with life testing experiments. Chechile[1] obtained the posterior distribution assuming that the random sample is taken from the Gumbel distribution using the conjugate prior. Corsini et al.[2] discussed the maximum likelihood (ML) algorithms and Cramer-Rao (CR) bounds for the location and scale parameters of the Gumbel distribution. Mousa[3] obtained the Bayesian estimation for the two parameters of the Gumbel distribution based on record values. Koutsoyiannis and Baloutsos[4] described that the Gumbel distribution has been the prevailing model for quantifying risk associated with extreme rainfall. Rasmussen and Gautam[5] extended the probability weighted moments (PWM) to what is called the generalized method of probability weighted moments (GPWM) because there is no reason why the PWMs provide the most efficient estimators of Gumbel parameters and quantile especially in hydrology. Malinowska and Szynal[6] obtained the Bayes estimators for the two parameters of a Gumbel distribution based on kth lower record values. Nadarajah and Kotz[7] introduced the beta Gumbel (BG) distribution and provided closed-form expressions for the moments, the asymptotic distribution of the extreme order statistics and discussed the maximum likelihood estimation procedure. Miladinovic and Tsokos[8] modified the classical Gumbel probability distribution in order to study the failure times of a given system. Park et al.[9] gave a novel equation for the scale parameter of the Gumbel distribution. Heo and Salas[10] examined the log-Gumbel distribution regarding quantile estimation and confidence intervals of quantiles. Thompson et al.[11] introduced a distributional hypothesis test for left censored Gumbel observations based on the probability plot correlation coefficient (PPCC).

The mixture models have received great attention of the analysts in the recent era. These models include finite and infinite number of components that can analyze different datasets. A finite mixture of probability distribution is suitable to study a population categorized in number of subpopulations. A population of lifetimes of certain electrical elements can be classified into number of subpopulations based on causes of failures. The analysis of mixture models under Bayesian framework has developed a significant interest among the statisticians. The authors dealing with Bayesian analysis of mixture models include Saleem and Aslam[12], Saleem et al.[13], Majeed and Aslam[14] and Kazmi et al.[15]. These contributions to the mixture models are the great motivations for the resent study.

We considered two component mixture of Gumbel type II distribution. The population of certain items is assumed to be partitioned into two subpopulations. The randomly selected observations from the said population are considered to be a part of one of the above mentioned subpopulations. These subpopulations are assumed to follow the Gumbel type II distribution. Therefore, the two components mixture of Gumbel type II distributions has been proposed to model this population. The observations have been assumed to be right censored. The inverse transformation technique of simulation under a probabilistic mixing has been used to generate data and to evaluate the performance of different estimators.

2. THE MODEL AND LIKELIHOOD FUNCTION

A density function for mixture of two components densities with mixing weights (p,q) is:

$$f(x) = pf_1(x) + qf_2(x)$$
$$0 < p < 1 \,,\; q = 1 - p \tag{1}$$

The following Gumbel type II distribution is considered for both mixture densities:

$$f(x_i; \alpha_i, \beta_i) = \alpha_i \beta_i x_i^{-(\alpha_i+1)} e^{-\beta_i x_i^{-\alpha_i}}$$
$$x_i > 0 ,\; \alpha_i > 0 ,\; \beta_i > 0 \;;\; i = 1,2 \tag{2}$$

With the cumulative distribution function as:

$$F(x_i; \alpha_i, \beta_i) = 1 - e^{-\beta_i x_i^{-\alpha_i}} \tag{3}$$

The cumulative distribution function for the mixture model is:

$$F(x) = pF_1(x) + qF_2(x) \tag{4}$$

Suppose n items are put on a life testing experiment and r units failed until time T while, n – r units are still working. Now based on causes of failure, the failed items are assumed to come either from subpopulation 1 or from subpopulation 2. Therefore it can be observed that r_1 and r_2 failed items come from first and second subpopulation respectively. Where $r = r_1 + r_2$. The remaining n – r items are assumed to be censored observations. The likelihood function for above censored data can be obtained as:

$$L(\beta_1, \beta_2, p | \underline{x}) = \prod_{j=1}^{r_1} \{ pf_1(x_{1j}) \}$$
$$\times \prod_{j=1}^{r_2} \{ pf_2(x_{2j}) \} [1 - F(t)]^{n-r} \tag{5}$$

$$L(\beta_1, \beta_2, p | \underline{x}) \propto \beta_1^{r_1} \beta_2^{r_2} p^{r_1} q^{r_2} e^{-\beta_1 \sum_{j=1}^{r_1} x_{1j} - \beta_2 \sum_{j=1}^{r_2} x_{2j}}$$
$$\times \left[pe^{-\beta_1 t^{-\alpha_1}} + qe^{-\beta_2 t^{-\alpha_2}} \right]^{n-r} \tag{6}$$

Expanding the last term by binomial expansion the likelihood function becomes:

$$L(\beta_1,\beta_2,p|\underline{x}) \propto \sum_{k=0}^{n-r}\binom{n-r}{k}\beta_1^{r_1}\beta_2^{r_2}p^{n-k-r_2}q^{r_2+k} \times e^{-\beta_1 A_{1k}}e^{-\beta_2 A_{2k}} \qquad (7)$$

Where $A_{1k}=\sum_{j=1}^{r_1}x_{1j}+(n-r-k)t^{-\alpha_1}$

and $A_{2k}=\sum_{j=1}^{r_2}x_{2j}+kt^{-\alpha_2}$

3. THE POSTERIOR ANALYSIS UNDER THE ASSUMPTION OF UNIFORM PRIOR

One of the most widely used non-informative priors, proposed by Laplace[16], is a uniform prior. It has been applied to many problems, and often the results are entirely satisfactory. This prior has been used for the posterior estimation.

Let $\beta_1 \in \text{Uniform } \forall\ \beta_1 \in (0,\infty)$, $\beta_2 \in \text{Uniform } \forall\ \beta_2 \in (0,\infty)$ and $p \sim U(0,1)$. Assuming independence, these priors result into a joint prior that is proportional to a constant. That joint prior has been used to derive the joint posterior distribution of $\beta_1, \beta_2 \text{ and } p$. The marginal distribution for each parameter can be obtained by integrating the joint posterior distribution with respect to nuisance parameters. The joint posterior distribution is:

$$p(\beta_1,\beta_2,p|\underline{x}) = \frac{1}{C_1}\sum_{k=0}^{n-r}\binom{n-r}{k}\beta_1^{r_1}\beta_2^{r_2}p^{n-k-r_2}q^{r_2+k} \times e^{-\beta_1 A_{1k}}e^{-\beta_2 A_{2k}} \qquad (8)$$

Where $C_1=\sum_{k=0}^{n-r}\binom{n-r}{k}B(\theta_{1k},\theta_{2k})\frac{\Gamma(r_1+1)}{(A_{1k})^{r_1+1}}\frac{\Gamma(r_2+1)}{(A_{2k})^{r_2+1}}$,

$$\theta_{1k}=n-k-r_2+1 \text{ and } \theta_{2k}=r_2+k+1$$

Using the posterior distribution, discussed in (8), the Bayes estimators and posterior risks under different loss functions have been derived and presented in the following.

Bayes estimators and associated risks under uniform prior using squared error loss function (SELF) are:

$$\beta_{1,SELF} = \frac{1}{C_1}\sum_{k=0}^{n-r}\binom{n-r}{k}B(\theta_{1k},\theta_{2k})\frac{\Gamma(r_1+2)}{(A_{1k})^{r_1+2}}\frac{\Gamma(r_2+1)}{(A_{2k})^{r_2+1}}$$

$$\rho(\beta_{1,SELF}) = \frac{1}{C_1}\sum_{k=0}^{n-r}\binom{n-r}{k}B(\theta_{1k},\theta_{2k})\frac{\Gamma(r_1+3)}{(A_{1k})^{r_1+3}}\frac{\Gamma(r_2+1)}{(A_{2k})^{r_2+1}}$$

$$-\left[\frac{1}{C_1}\sum_{k=0}^{n-r}\binom{n-r}{k}B(\theta_{1k},\theta_{2k})\frac{\Gamma(r_1+2)}{(A_{1k})^{r_1+2}}\frac{\Gamma(r_2+1)}{(A_{2k})^{r_2+1}}\right]^2$$

$$\beta_{2,SELF} = \frac{1}{C_1}\sum_{k=0}^{n-r}\binom{n-r}{k}B(\theta_{1k},\theta_{2k})\frac{\Gamma(r_1+1)}{(A_{1k})^{r_1+1}}\frac{\Gamma(r_2+2)}{(A_{2k})^{r_2+2}}$$

$$\rho(\beta_{2,SELF}) = \frac{1}{C_1}\sum_{k=0}^{n-r}\binom{n-r}{k}B(\theta_{1k},\theta_{2k})\frac{\Gamma(r_1+1)}{(A_{1k})^{r_1+1}}\frac{\Gamma(r_2+3)}{(A_{2k})^{r_2+3}}$$

$$-\left[\frac{1}{C_1}\sum_{k=0}^{n-r}\binom{n-r}{k}B(\theta_{1k},\theta_{2k})\frac{\Gamma(r_1+1)}{(A_{1k})^{r_1+1}}\frac{\Gamma(r_2+2)}{(A_{2k})^{r_2+2}}\right]^2$$

$$p_{SELF} = \frac{1}{C_1}\sum_{k=0}^{n-r}\binom{n-r}{k}B(\theta_{1k}+1,\theta_{2k})\frac{\Gamma(r_1+1)}{(A_{1k})^{r_1+1}}\frac{\Gamma(r_2+1)}{(A_{2k})^{r_2+1}}$$

$$\rho(p_{SELF}) = \frac{1}{C_1}\sum_{k=0}^{n-r}\binom{n-r}{k}B(\theta_{1k}+2,\theta_{2k})\frac{\Gamma(r_1+1)}{(A_{1k})^{r_1+1}}\frac{\Gamma(r_2+1)}{(A_{2k})^{r_2+1}}$$

$$-\left[\frac{1}{C_1}\sum_{k=0}^{n-r}\binom{n-r}{k}B(\theta_{1k}+1,\theta_{2k})\frac{\Gamma(r_1+1)}{(A_{1k})^{r_1+1}}\frac{\Gamma(r_2+1)}{(A_{2k})^{r_2+1}}\right]^2$$

Bayes estimators and risk under uniform prior using quadratic loss function (QLF) are:

$$\beta_{1,QLF}=\frac{\sum_{k=0}^{n-r}\binom{n-r}{k}B(\theta_{1k},\theta_{2k})\frac{\Gamma(r_1)}{(A_{1k})^{r_1}}\frac{\Gamma(r_2+1)}{(A_{2k})^{r_2+1}}}{\sum_{k=0}^{n-r}\binom{n-r}{k}B(\theta_{1k},\theta_{2k})\frac{\Gamma(r_1-1)}{(A_{1k})^{r_1-1}}\frac{\Gamma(r_2+1)}{(A_{2k})^{r_2+1}}}$$

$$\rho(\beta_{1,QLF})=1-\frac{\frac{1}{C_1}\left[\sum_{k=0}^{n-r}\binom{n-r}{k}B(\theta_{1k},\theta_{2k})\frac{\Gamma(r_1)}{(A_{1k})^{r_1}}\frac{\Gamma(r_2+1)}{(A_{2k})^{r_2+1}}\right]^2}{\sum_{k=0}^{n-r}\binom{n-r}{k}B(\theta_{1k},\theta_{2k})\frac{\Gamma(r_1-1)}{(A_{1k})^{r_1-1}}\frac{\Gamma(r_2+1)}{(A_{2k})^{r_2+1}}}$$

$$\beta_{2,QLF}=\frac{\sum_{k=0}^{n-r}\binom{n-r}{k}B(\theta_{1k},\theta_{2k})\frac{\Gamma(r_1+1)}{(A_{1k})^{r_1+1}}\frac{\Gamma(r_2)}{(A_{2k})^{r_2}}}{\sum_{k=0}^{n-r}\binom{n-r}{k}B(\theta_{1k},\theta_{2k})\frac{\Gamma(r_1+1)}{(A_{1k})^{r_1+1}}\frac{\Gamma(r_2-1)}{(A_{2k})^{r_2-1}}}$$

$$\rho(\beta_{2,QLF})=1-\frac{\frac{1}{C_1}\left[\sum_{k=0}^{n-r}\binom{n-r}{k}B(\theta_{1k},\theta_{2k})\frac{\Gamma(r_1+1)}{(A_{1k})^{r_1+1}}\frac{\Gamma(r_2)}{(A_{2k})^{r_2}}\right]^2}{\sum_{k=0}^{n-r}\binom{n-r}{k}B(\theta_{1k},\theta_{2k})\frac{\Gamma(r_1+1)}{(A_{1k})^{r_1+1}}\frac{\Gamma(r_2-1)}{(A_{2k})^{r_2-1}}}$$

$$p_{QLF}=\frac{\sum_{k=0}^{n-r}\binom{n-r}{k}B(\theta_{1k}-1,\theta_{2k})\frac{\Gamma(r_1+1)}{(A_{1k})^{r_1+1}}\frac{\Gamma(r_2+1)}{(A_{2k})^{r_2+1}}}{\sum_{k=0}^{n-r}\binom{n-r}{k}B(\theta_{1k}-2,\theta_{2k})\frac{\Gamma(r_1+1)}{(A_{1k})^{r_1+1}}\frac{\Gamma(r_2+1)}{(A_{2k})^{r_2+1}}}$$

$$\rho(p_{QLF})=1-\frac{\frac{1}{C_1}\left[\sum_{k=0}^{n-r}\binom{n-r}{k}B(\theta_{1k}-1,\theta_{2k})\frac{\Gamma(r_1+1)}{(A_{1k})^{r_1+1}}\frac{\Gamma(r_2+1)}{(A_{2k})^{r_2+1}}\right]^2}{\sum_{k=0}^{n-r}\binom{n-r}{k}B(\theta_{1k}-2,\theta_{2k})\frac{\Gamma(r_1+1)}{(A_{1k})^{r_1+1}}\frac{\Gamma(r_2+1)}{(A_{2k})^{r_2+1}}}$$

Bayes estimators and risk under uniform prior using weighted loss function (WLF) are:

$$\beta_{1,WLF} = \left[\frac{1}{C_1}\sum_{k=0}^{n-r}\binom{n-r}{k}B(\theta_{1k},\theta_{2k})\frac{\Gamma(r_1)}{(A_{1k})^{r_1}}\frac{\Gamma(r_2+1)}{(A_{2k})^{r_2+1}}\right]^{-1}$$

$$\rho(\beta_{1,WLF}) = \frac{1}{C_1}\sum_{k=0}^{n-r}\binom{n-r}{k}B(\theta_{1k},\theta_{2k})\frac{\Gamma(r_1+2)}{(A_{1k})^{r_1+2}}\frac{\Gamma(r_2+1)}{(A_{2k})^{r_2+1}}$$
$$-\left[\frac{1}{C_1}\sum_{k=0}^{n-r}\binom{n-r}{k}B(\theta_{1k},\theta_{2k})\frac{\Gamma(r_1)}{(A_{1k})^{r_1}}\frac{\Gamma(r_2+1)}{(A_{2k})^{r_2+1}}\right]^{-1}$$

$$\beta_{2,WLF} = \left[\frac{1}{C_1}\sum_{k=0}^{n-r}\binom{n-r}{k}B(\theta_{1k},\theta_{2k})\frac{\Gamma(r_1+1)}{(A_{1k})^{r_1+1}}\frac{\Gamma(r_2)}{(A_{2k})^{r_2}}\right]^{-1}$$

$$\rho(\beta_{2,WLF}) = \frac{1}{C_1}\sum_{k=0}^{n-r}\binom{n-r}{k}B(\theta_{1k},\theta_{2k})\frac{\Gamma(r_1+1)}{(A_{1k})^{r_1+1}}\frac{\Gamma(r_2+2)}{(A_{2k})^{r_2+2}}$$
$$-\left[\frac{1}{C_1}\sum_{k=0}^{n-r}\binom{n-r}{k}B(\theta_{1k},\theta_{2k})\frac{\Gamma(r_1+1)}{(A_{1k})^{r_1+1}}\frac{\Gamma(r_2)}{(A_{2k})^{r_2}}\right]^{-1}$$

$$p_{WLF} = \left[\frac{1}{C_1}\sum_{k=0}^{n-r}\binom{n-r}{k}B(\theta_{1k}-1,\theta_{2k})\frac{\Gamma(r_1+1)}{(A_{1k})^{r_1+1}}\frac{\Gamma(r_2+1)}{(A_{2k})^{r_2+1}}\right]^{-1}$$

$$\rho(p_{WLF}) = \frac{1}{C_1}\sum_{k=0}^{n-r}\binom{n-r}{k}B(\theta_{1k}+1,\theta_{2k})\frac{\Gamma(r_1+1)}{(A_{1k})^{r_1+1}}\frac{\Gamma(r_2+1)}{(A_{2k})^{r_2+1}}$$
$$-\left[\frac{1}{C_1}\sum_{k=0}^{n-r}\binom{n-r}{k}B(\theta_{1k}-1,\theta_{2k})\frac{\Gamma(r_1+1)}{(A_{1k})^{r_1+1}}\frac{\Gamma(r_2+1)}{(A_{2k})^{r_2+1}}\right]^{-1}$$

Bayes estimators and risk under uniform prior using precautionary loss function (PLF) are:

$$\beta_{1,PLF} = \left[\frac{1}{C_1}\sum_{k=0}^{n-r}(-1)^k\binom{n-r}{k}B(\theta_{1k},\theta_{2k})\frac{\Gamma(r_1+3)}{(A_{1k})^{r_1+3}}\frac{\Gamma(r_2+1)}{(A_{2k})^{r_2+1}}\right]^{\frac{1}{2}}$$

$$\rho(\beta_{1,PLF}) = 2\left[\frac{1}{C_1}\sum_{k=0}^{n-r}(-1)^k\binom{n-r}{k}B(\theta_{1k},\theta_{2k})\frac{\Gamma(r_1+3)}{(A_{1k})^{r_1+3}}\frac{\Gamma(r_2+1)}{(A_{2k})^{r_2+1}}\right]^{\frac{1}{2}}$$
$$-\frac{2}{C_1}\sum_{k=0}^{n-r}(-1)^k\binom{n-r}{k}B(\theta_{1k},\theta_{2k})\frac{\Gamma(r_1+2)}{(A_{1k})^{r_1+2}}\frac{\Gamma(r_2+1)}{(A_{2k})^{r_2+1}}$$

$$\beta_{2,PLF} = \left[\frac{1}{C_1}\sum_{k=0}^{n-r}\binom{n-r}{k}B(\theta_{1k},\theta_{2k})\frac{\Gamma(r_1+1)}{(A_{1k})^{r_1+1}}\frac{\Gamma(r_2+3)}{(A_{2k})^{r_2+3}}\right]^{\frac{1}{2}}$$

$$\rho(\beta_{2,PLF}) = 2\left[\frac{1}{C_1}\sum_{k=0}^{n-r}\binom{n-r}{k}B(\theta_{1k},\theta_{2k})\frac{\Gamma(r_1+1)}{(A_{1k})^{r_1+1}}\frac{\Gamma(r_2+3)}{(A_{2k})^{r_2+3}}\right]^{\frac{1}{2}}$$
$$-\frac{2}{C_1}\sum_{k=0}^{n-r}\binom{n-r}{k}B(\theta_{1k},\theta_{2k})\frac{\Gamma(r_1+1)}{(A_{1k})^{r_1+1}}\frac{\Gamma(r_2+2)}{(A_{2k})^{r_2+2}}$$

$$p_{PLF} = \left[\frac{1}{C_1}\sum_{k=0}^{n-r}\binom{n-r}{k}B(\theta_{1k}+2,\theta_{2k})\frac{\Gamma(r_1+1)}{(A_{1k})^{r_1+1}}\frac{\Gamma(r_2+1)}{(A_{2k})^{r_2+1}}\right]^{\frac{1}{2}}$$

$$\rho(p_{PLF}) = 2\left[\frac{1}{C_1}\sum_{k=0}^{n-r}\binom{n-r}{k}B(\theta_{1k}+2,\theta_{2k})\frac{\Gamma(r_1+1)}{(A_{1k})^{r_1+1}}\frac{\Gamma(r_2+1)}{(A_{2k})^{r_2+1}}\right]^{\frac{1}{2}} - \frac{2}{C_1}\sum_{k=0}^{n-r}\binom{n-r}{k}B(\theta_{1k}+1,\theta_{2k})\frac{\Gamma(r_1+1)}{(A_{1k})^{r_1+1}}\frac{\Gamma(r_2+1)}{(A_{2k})^{r_2+1}}$$

4. THE POSTERIOR ANALYSIS UNDER THE ASSUMPTION OF JEFFREYS PRIOR

Another non-informative prior has been suggested by Jeffreys[17] which is frequently used in situations where one does not have much information about the parameters. This is defined as the distribution of the parameters proportional to the square root of the determinants of the Fisher information matrix i.e.

$$p(\underline{\beta}) \propto \left\{\left|I(\underline{\beta})\right|\right\}^{\frac{1}{2}}$$

Where $\underline{\beta} = (\beta_1, \beta_2)$ is the vector of parameters and $I(\underline{\beta})$ is (2×2) Fisher information matrix, defined as;

$$I(\underline{\beta}) = -E\left[\frac{\partial^2 \log f_i(x_i|\beta_i)}{\partial \beta_i^2}\right] \quad i = 1,2$$

Where $f_i(x_i|\beta_i)$ have been defined in (2) and $p \sim U(0,1)$. Assuming independence, the joint prior is obtained as:

$$h(\beta_1, \beta_2, p) \propto \frac{1}{\beta_1 \beta_2}. \tag{9}$$

The joint posterior distribution using the above prior is:

$$p(\beta_1, \beta_2, p|\underline{x}) = \frac{1}{C_2}\sum_{k=0}^{n-r}\binom{n-r}{k}\beta_1^{r_1-1}\beta_2^{r_2-1}p^{n-k-r_2}q^{r_2+k} \times e^{-\beta_1 A_{1k}}e^{-\beta_2 A_{2k}} \tag{10}$$

Where $C_2 = \sum_{k=0}^{n-r} \binom{n-r}{k} B\left(\theta_{1k}, \theta_{2k}\right) \frac{\Gamma(r_1)}{\left(A_{1k}\right)^{r_1}} \frac{\Gamma(r_2)}{\left(A_{2k}\right)^{r_2}}$

The Bayes estimators and associated posterior risks have been derived under different loss functions using the posterior distribution (10).

Bayes estimators and associated risks under Jeffreys prior using squared error loss function (SELF) are:

$$\beta_{1,SELF} = \frac{1}{C_2} \sum_{k=0}^{n-r} \binom{n-r}{k} B\left(\theta_{1k}, \theta_{2k}\right) \frac{\Gamma(r_1+1)}{\left(A_{1k}\right)^{r_1+1}} \frac{\Gamma(r_2)}{\left(A_{2k}\right)^{r_2}}$$

$$\rho\left(\beta_{1,SELF}\right) = \frac{1}{C_2} \sum_{k=0}^{n-r} \binom{n-r}{k} B\left(\theta_{1k}, \theta_{2k}\right) \frac{\Gamma(r_1+2)}{\left(A_{1k}\right)^{r_1+2}} \frac{\Gamma(r_2)}{\left(A_{2k}\right)^{r_2}}$$

$$- \left[\frac{1}{C_2} \sum_{k=0}^{n-r} \binom{n-r}{k} B\left(\theta_{1k}, \theta_{2k}\right) \frac{\Gamma(r_1+1)}{\left(A_{1k}\right)^{r_1+1}} \frac{\Gamma(r_2)}{\left(A_{2k}\right)^{r_2}}\right]^2$$

$$\beta_{2,SELF} = \frac{1}{C_2} \sum_{k=0}^{n-r} \binom{n-r}{k} B\left(\theta_{1k}, \theta_{2k}\right) \frac{\Gamma(r_1)}{\left(A_{1k}\right)^{r_1}} \frac{\Gamma(r_2+1)}{\left(A_{2k}\right)^{r_2+1}}$$

$$\rho\left(\beta_{2,SELF}\right) = \frac{1}{C_2} \sum_{k=0}^{n-r} \binom{n-r}{k} B\left(\theta_{1k}, \theta_{2k}\right) \frac{\Gamma(r_1)}{\left(A_{1k}\right)^{r_1}} \frac{\Gamma(r_2+2)}{\left(A_{2k}\right)^{r_2+2}}$$

$$- \left[\frac{1}{C_2} \sum_{k=0}^{n-r} \binom{n-r}{k} B\left(\theta_{1k}, \theta_{2k}\right) \frac{\Gamma(r_1)}{\left(A_{1k}\right)^{r_1}} \frac{\Gamma(r_2+1)}{\left(A_{2k}\right)^{r_2+1}}\right]^2$$

$$p_{SELF} = \frac{1}{C_2} \sum_{k=0}^{n-r} \binom{n-r}{k} B\left(\theta_{1k}+1, \theta_{2k}\right) \frac{\Gamma(r_1)}{\left(A_{1k}\right)^{r_1}} \frac{\Gamma(r_2)}{\left(A_{2k}\right)^{r_2}}$$

$$\rho\left(p_{SELF}\right) = \frac{1}{C_2} \sum_{k=0}^{n-r} \binom{n-r}{k} B\left(\theta_{1k}+2, \theta_{2k}\right) \frac{\Gamma(r_1)}{\left(A_{1k}\right)^{r_1}} \frac{\Gamma(r_2)}{\left(A_{2k}\right)^{r_2}}$$

$$- \left[\frac{1}{C_2} \sum_{k=0}^{n-r} \binom{n-r}{k} B\left(\theta_{1k}+1, \theta_{2k}\right) \frac{\Gamma(r_1)}{\left(A_{1k}\right)^{r_1}} \frac{\Gamma(r_2)}{\left(A_{2k}\right)^{r_2}}\right]^2$$

Bayes estimators and risk under Jeffreys prior using quadratic loss function (QLF) are:

$$\beta_{1,QLF}=\frac{\sum_{k=0}^{n-r}\binom{n-r}{k}B(\theta_{1k},\theta_{2k})\frac{\Gamma(r_1-1)}{(A_{1k})^{r_1-1}}\frac{\Gamma(r_2)}{(A_{2k})^{r_2}}}{\sum_{k=0}^{n-r}\binom{n-r}{k}B(\theta_{1k},\theta_{2k})\frac{\Gamma(r_1-2)}{(A_{1k})^{r_1-2}}\frac{\Gamma(r_2)}{(A_{2k})^{r_2}}}$$

$$\rho(\beta_{1,QLF})=1-\frac{\frac{1}{C_2}\left[\sum_{k=0}^{n-r}\binom{n-r}{k}B(\theta_{1k},\theta_{2k})\frac{\Gamma(r_1-1)}{(A_{1k})^{r_1-1}}\frac{\Gamma(r_2)}{(A_{2k})^{r_2}}\right]^2}{\sum_{k=0}^{n-r}\binom{n-r}{k}B(\theta_{1k},\theta_{2k})\frac{\Gamma(r_1-2)}{(A_{1k})^{r_1-2}}\frac{\Gamma(r_2)}{(A_{2k})^{r_2}}}$$

$$\beta_{2,QLF}=\frac{\sum_{k=0}^{n-r}\binom{n-r}{k}B(\theta_{1k},\theta_{2k})\frac{\Gamma(r_1)}{(A_{1k})^{r_1}}\frac{\Gamma(r_2-1)}{(A_{2k})^{r_2-1}}}{\sum_{k=0}^{n-r}\binom{n-r}{k}B(\theta_{1k},\theta_{2k})\frac{\Gamma(r_1)}{(A_{1k})^{r_1}}\frac{\Gamma(r_2-2)}{(A_{2k})^{r_2-2}}}$$

$$\rho(\beta_{2,QLF})=1-\frac{\frac{1}{C_2}\left[\sum_{k=0}^{n-r}\binom{n-r}{k}B(\theta_{1k},\theta_{2k})\frac{\Gamma(r_1)}{(A_{1k})^{r_1}}\frac{\Gamma(r_2-1)}{(A_{2k})^{r_2-1}}\right]^2}{\sum_{k=0}^{n-r}\binom{n-r}{k}B(\theta_{1k},\theta_{2k})\frac{\Gamma(r_1)}{(A_{1k})^{r_1}}\frac{\Gamma(r_2-2)}{(A_{2k})^{r_2-2}}}$$

$$p_{QLF}=\frac{\sum_{k=0}^{n-r}\binom{n-r}{k}B(\theta_{1k}-1,\theta_{2k})\frac{\Gamma(r_1)}{(A_{1k})^{r_1}}\frac{\Gamma(r_2)}{(A_{2k})^{r_2}}}{\sum_{k=0}^{n-r}\binom{n-r}{k}B(\theta_{1k}-2,\theta_{2k})\frac{\Gamma(r_1)}{(A_{1k})^{r_1}}\frac{\Gamma(r_2)}{(A_{2k})^{r_2}}}$$

$$\rho(p_{QLF})=1-\frac{\frac{1}{C_2}\left[\sum_{k=0}^{n-r}\binom{n-r}{k}B(\theta_{1k}-1,\theta_{2k})\frac{\Gamma(r_1)}{(A_{1k})^{r_1}}\frac{\Gamma(r_2)}{(A_{2k})^{r_2}}\right]^2}{\sum_{k=0}^{n-r}\binom{n-r}{k}B(\theta_{1k}-2,\theta_{2k})\frac{\Gamma(r_1)}{(A_{1k})^{r_1}}\frac{\Gamma(r_2)}{(A_{2k})^{r_2}}}$$

Bayes estimators and risk under Jeffreys prior using weighted loss function (WLF) are:

$$\beta_{1,WLF} = \left[\frac{1}{C_2} \sum_{k=0}^{n-r} \binom{n-r}{k} B(\theta_{1k}, \theta_{2k}) \frac{\Gamma(r_1 - 1)}{(A_{1k})^{r_1 - 1}} \frac{\Gamma(r_2)}{(A_{2k})^{r_2}} \right]^{-1}$$

$$\rho(\beta_{1,WLF}) = \frac{1}{C_2} \sum_{k=0}^{n-r} \binom{n-r}{k} B(\theta_{1k}, \theta_{2k}) \frac{\Gamma(r_1 + 1)}{(A_{1k})^{r_1 + 1}} \frac{\Gamma(r_2)}{(A_{2k})^{r_2}} - \left[\frac{1}{C_2} \sum_{k=0}^{n-r} \binom{n-r}{k} B(\theta_{1k}, \theta_{2k}) \frac{\Gamma(r_1 - 1)}{(A_{1k})^{r_1 - 1}} \frac{\Gamma(r_2)}{(A_{2k})^{r_2}} \right]^{-1}$$

$$\beta_{2,WLF} = \left[\frac{1}{C_2} \sum_{k=0}^{n-r} \binom{n-r}{k} B(\theta_{1k}, \theta_{2k}) \frac{\Gamma(r_1)}{(A_{1k})^{r_1}} \frac{\Gamma(r_2 - 1)}{(A_{2k})^{r_2 - 1}} \right]^{-1}$$

$$\rho(\beta_{2,WLF}) = \frac{1}{C_2} \sum_{k=0}^{n-r} \binom{n-r}{k} B(\theta_{1k}, \theta_{2k}) \frac{\Gamma(r_1)}{(A_{1k})^{r_1}} \frac{\Gamma(r_2 + 1)}{(A_{2k})^{r_2 + 1}} - \left[\frac{1}{C_2} \sum_{k=0}^{n-r} \binom{n-r}{k} B(\theta_{1k}, \theta_{2k}) \frac{\Gamma(r_1)}{(A_{1k})^{r_1}} \frac{\Gamma(r_2 - 1)}{(A_{2k})^{r_2 - 1}} \right]^{-1}$$

$$p_{WLF} = \left[\frac{1}{C_2} \sum_{k=0}^{n-r} \binom{n-r}{k} B(\theta_{1k} - 1, \theta_{2k}) \frac{\Gamma(r_1)}{(A_{1k})^{r_1}} \frac{\Gamma(r_2)}{(A_{2k})^{r_2}} \right]^{-1}$$

$$\rho(p_{WLF}) = \frac{1}{C_2} \sum_{k=0}^{n-r} \binom{n-r}{k} B(\theta_{1k} + 1, \theta_{2k}) \frac{\Gamma(r_1)}{(A_{1k})^{r_1}} \frac{\Gamma(r_2)}{(A_{2k})^{r_2}} - \left[\frac{1}{C_2} \sum_{k=0}^{n-r} \binom{n-r}{k} B(\theta_{1k} - 1, \theta_{2k}) \frac{\Gamma(r_1)}{(A_{1k})^{r_1}} \frac{\Gamma(r_2)}{(A_{2k})^{r_2}} \right]^{-1}$$

Bayes estimators and risk under Jeffreys prior using precautionary loss function (PLF) are:

$$\beta_{1,PLF} = \left[\frac{1}{C_2} \sum_{k=0}^{n-r} \binom{n-r}{k} B(\theta_{1k}, \theta_{2k}) \frac{\Gamma(r_1 + 2)}{(A_{1k})^{r_1 + 2}} \frac{\Gamma(r_2)}{(A_{2k})^{r_2}} \right]^{\frac{1}{2}}$$

$$\rho(\beta_{1,PLF}) = 2\left[\frac{1}{C_2} \sum_{k=0}^{n-r} \binom{n-r}{k} B(\theta_{1k}, \theta_{2k}) \frac{\Gamma(r_1 + 2)}{(A_{1k})^{r_1 + 2}} \frac{\Gamma(r_2)}{(A_{2k})^{r_2}} \right]^{\frac{1}{2}} - \frac{2}{C_2} \sum_{k=0}^{n-r} \binom{n-r}{k} B(\theta_{1k}, \theta_{2k}) \frac{\Gamma(r_1 + 1)}{(A_{1k})^{r_1 + 1}} \frac{\Gamma(r_2)}{(A_{2k})^{r_2}}$$

$$\beta_{2,PLF}=\left[\frac{1}{C_2}\sum_{k=0}^{n-r}\binom{n-r}{k}B(\theta_{1k},\theta_{2k})\frac{\Gamma(r_1)}{(A_{1k})^{r_1}}\frac{\Gamma(r_2+2)}{(A_{2k})^{r_2+2}}\right]^{\frac{1}{2}}$$

$$\rho(\beta_{2,PLF})=2\left[\frac{1}{C_2}\sum_{k=0}^{n-r}\binom{n-r}{k}B(\theta_{1k},\theta_{2k})\frac{\Gamma(r_1)}{(A_{1k})^{r_1}}\frac{\Gamma(r_2+2)}{(A_{2k})^{r_2+2}}\right]^{\frac{1}{2}}$$
$$-\frac{2}{C_2}\sum_{k=0}^{n-r}\binom{n-r}{k}B(\theta_{1k},\theta_{2k})\frac{\Gamma(r_1)}{(A_{1k})^{r_1}}\frac{\Gamma(r_2+1)}{(A_{2k})^{r_2+1}}$$

$$p_{PLF}=\left[\frac{1}{C_2}\sum_{k=0}^{n-r}\binom{n-r}{k}B(\theta_{1k}+2,\theta_{2k})\frac{\Gamma(r_1)}{(A_{1k})^{r_1}}\frac{\Gamma(r_2)}{(A_{2k})^{r_2}}\right]^{\frac{1}{2}}$$

$$\rho(p_{PLF})=2\left[\frac{1}{C_2}\sum_{k=0}^{n-r}\binom{n-r}{k}B(\theta_{1k}+2,\theta_{2k})\frac{\Gamma(r_1)}{(A_{1k})^{r_1}}\frac{\Gamma(r_2)}{(A_{2k})^{r_2}}\right]^{\frac{1}{2}}$$
$$-\frac{2}{C_2}\sum_{k=0}^{n-r}\binom{n-r}{k}B(\theta_{1k}+1,\theta_{2k})\frac{\Gamma(r_1)}{(A_{1k})^{r_1}}\frac{\Gamma(r_2)}{(A_{2k})^{r_2}}$$

5. POSTERIOR PREDICTIVE DISTRIBUTIONS AND INTERVALS

The posterior predictive distributions under uniform and Jeffreys priors are:

$$p(y|\underline{x})=\frac{1}{C_1}\sum_{k=0}^{n-r}\binom{n-r}{k}\left[\frac{B(\theta_{1k}+1,\theta_{2k})\Gamma(r_1+2)\Gamma(r_2+1)\alpha_1 y^{-(\alpha_1+1)}}{\left(A_{1k}+y^{-\alpha_1}\right)^{r_1+2}(A_{2k})^{r_2+1}}\right.$$
$$\left.+\frac{B(\theta_{1k},\theta_{2k}+1)\Gamma(r_1+1)\Gamma(r_2+2)\alpha_2 y^{-(\alpha_2+1)}}{(A_{1k})^{r_1+1}\left(A_{2k}+y^{-\alpha_2}\right)^{r_2+2}}\right]$$

$$p(y|\underline{x})=\frac{1}{C_2}\sum_{k=0}^{n-r}\binom{n-r}{k}\left[\frac{B(\theta_{1k}+1,\theta_{2k})\Gamma(r_1+1)\Gamma(r_2)\alpha_1 y^{-(\alpha_1+1)}}{\left(A_{1k}+y^{-\alpha_1}\right)^{r_1+1}(A_{2k})^{r_2}}\right.$$
$$\left.+\frac{B(\theta_{1k},\theta_{2k}+1)\Gamma(r_1)\Gamma(r_2+1)\alpha_2 y^{-(\alpha_2+1)}}{(A_{1k})^{r_1}\left(A_{2k}+y^{-\alpha_2}\right)^{r_2+1}}\right]$$

The posterior predictive intervals under uniform prior can be obtained by solving the following two equations respectively.

$$\frac{k}{2}=\frac{1}{C_1}\sum_{k=0}^{n-r}\binom{n-r}{k}$$

$$\left[\frac{B(\theta_{1k}+1,\theta_{2k})\Gamma(r_1+1)\Gamma(r_2+1)}{(A_{2k})^{r_2+1}}\left\{\frac{1}{(A_{1k})^{r_1+1}}-\frac{1}{\left(A_{1k}+L^{-\alpha_1}\right)^{r_1+1}}\right\}\right.$$

$$\left.+\frac{B(\theta_{1k},\theta_{2k}+1)\Gamma(r_1+1)\Gamma(r_2+1)}{(A_{1k})^{r_1+1}}\left\{\frac{1}{(A_{2k})^{r_2+1}}-\frac{1}{\left(A_{2k}+L^{-\alpha_2}\right)^{r_2+1}}\right\}\right]$$

$$1-\frac{k}{2}=\frac{1}{C_1}\sum_{k=0}^{n-r}\binom{n-r}{k}$$

$$\left[\frac{B(\theta_{1k}+1,\theta_{2k})\Gamma(r_1+1)\Gamma(r_2+1)}{(A_{2k})^{r_2+1}}\left\{\frac{1}{(A_{1k})^{r_1+1}}-\frac{1}{\left(A_{1k}+U^{-\alpha_1}\right)^{\gamma+1}}\right\}\right.$$

$$\left.+\frac{B(\theta_{1k},\theta_{2k}+1)\Gamma(r_1+1)\Gamma(r_2+1)}{(A_{1k})^{r_1+1}}\left\{\frac{1}{(A_{2k})^{r_2+1}}-\frac{1}{\left(A_{2k}+U^{-\alpha_2}\right)^{r_2+1}}\right\}\right]$$

Where 'k' is level of significance

The posterior predictive intervals under uniform prior can be obtained by solving the following two equations respectively.

$$\frac{k}{2}=\frac{1}{C_1}\sum_{k=0}^{n-r}\binom{n-r}{k}$$

$$\left[\frac{B(\theta_{1k}+1,\theta_{2k})\Gamma(r_1)\Gamma(r_2)}{(A_{2k})^{r_2}}\left\{\frac{1}{(A_{1k})^{r_1}}-\frac{1}{\left(A_{1k}+L^{-\alpha_1}\right)^{r_1}}\right\}\right.$$

$$\left.+\frac{B(\theta_{1k},\theta_{2k}+1)\Gamma(r_1)\Gamma(r_2)}{(A_{1k})^{r_1}}\left\{\frac{1}{(A_{2k})^{r_2}}-\frac{1}{\left(A_{2k}+L^{-\alpha_2}\right)^{r_2}}\right\}\right]$$

$$1-\frac{k}{2}=\frac{1}{C_1}\sum_{k=0}^{n-r}\binom{n-r}{k}$$

$$\left[\frac{B(\theta_{1k}+1,\theta_{2k})\Gamma(r_1)\Gamma(r_2)}{(A_{2k})^{r_2}}\left\{\frac{1}{(A_{1k})^{r_1}}-\frac{1}{\left(A_{1k}+U^{-\alpha_1}\right)^{r_1}}\right\}\right.$$

$$\left.+\frac{B(\theta_{1k},\theta_{2k}+1)\Gamma(r_1)\Gamma(r_2)}{(A_{1k})^{r_1+1}}\left\{\frac{1}{(A_{2k})^{r_2}}-\frac{1}{\left(A_{2k}+U^{-\alpha_2}\right)^{r_2}}\right\}\right]$$

6. SIMULATION STUDY

A simulation study has been conducted to assess and compare the performance of Bayes estimators and to analyse the impact of sample size, mixing weight and magnitude of parametric values on the Bayes estimators. Samples of sizes n = 100, 200, 300, 400 and 500 have been generated by inverse transformation method from two components mixture of Gumbel type II distribution. The parametric values used are: $(\beta_1, \beta_2) \in \{(3,6),(6,9),(10,12)\}$ and $p \in (0.30, 0.45)$. The probabilistic mixing has been used to generate the mixture data. For each observation a random number has been generated from $U(0,1)$. If $u < p$ the observation has been randomly taken from first subpopulation and if $u > p$ then the observation have been taken from the second subpopulation.

The observations above a fixed censoring time T have been assumed to be right censored. Under each combination of parametric values, the choice of censoring time has been made so that the censoring rate in the respective sample has been 20%. As one sample cannot completely describe the behaviour and properties of the Bayes estimators, the results have been replicated 1000 times and the average of results has been presented in the tables below.

The simulation study indicates that by increasing the sample size the estimated values each parameter converges to the true value and the magnitude of corresponding posterior risks decreases. However, the posterior risks seem to be quite large for relatively larger values of the parameters. It can also be observed that all the parameters are over estimated with few exceptions. The extend of over estimation is more intensive for larger values of the parameters. This indicates that the posterior distributions are positively skewed. In comparison of priors, the performance of estimates under the Jeffreys prior seems better than those under uniform prior for each loss function and mixing weight. While, in case of loss functions, the estimates under quadratic loss function (QLF) are associated with the minimum risks. It is also interesting to note that the amount of risks under precautionary loss function (PLF) and weighted loss function (WLF) are converging to each other with increase in sample size. The posterior risks under QLF are approximately half of the risks under WLF and PLF. The increase in the value of mixing proportion p imposes positive impact on the performance of estimators of β_1 and it negatively affects the performance of the estimators of β_2 and vice versa. This is because the increasing value of p tend to increase the value of r_1 (the number of observations selected from the Gumbel type II distribution having parameter β_1), which will result in lesser posterior risks. The increasing values of the main parameters (β_1 and β_2) are having a negative effect on the behavior of the estimators of mixing proportion p. The expressions for complete samples can simply be obtained by increasing the test termination time to unity. The risks associated with estimates under complete samples are expected to reduce as no information will be lost then.

Table 1. Bayes estimates and posterior risks under uniform prior using β1= 3, β2 = 6 and p = 0.30

n	$\beta_1=3$			
	SELF	QLF	WLF	PLF
100	3.28620 0.05344	3.22048 0.00270	3.25301 0.01195	3.30310 0.01216
200	3.26174 0.02592	3.22912 0.00135	3.24535 0.00590	3.27001 0.00595
300	3.19922 0.01237	3.18322 0.00068	3.19120 0.00289	3.20324 0.00290
400	3.10967 0.00777	3.09931 0.00045	3.10448 0.00187	3.11228 0.00187
500	3.03695 0.00555	3.02936 0.00034	3.03315 0.00137	3.03886 0.00137
n	$\beta_2=6$			
	SELF	QLF	WLF	PLF
100	6.56364 0.06136	6.43237 0.00310	6.49734 0.01372	6.59739 0.01397
200	6.51478 0.02976	6.44963 0.00155	6.48204 0.00677	6.53130 0.00683

300	6.38990 0.01421	6.35795 0.00078	6.37389 0.00331	6.39794 0.00333
400	6.21105 0.00893	6.19035 0.00052	6.20068 0.00215	6.21625 0.00215
500	6.06581 0.00638	6.05064 0.00039	6.05821 0.00157	6.06961 0.00157
n	p = 0.30			
	SELF	QLF	WLF	PLF
100	0.33300 0.00297	0.32634 0.00015	0.32964 0.00066	0.33471 0.00068
200	0.33052 0.00144	0.32722 0.00008	0.32886 0.00033	0.33136 0.00033
300	0.32419 0.00069	0.32257 0.00004	0.32337 0.00016	0.32460 0.00016
400	0.31511 0.00043	0.31406 0.00003	0.31459 0.00010	0.31538 0.00010
500	0.30774 0.00031	0.30698 0.00002	0.30736 0.00008	0.30794 0.00008

Table 2. Bayes estimates and posterior risks under uniform prior using β1 = 3, β2 = 6 and p = 0.45 $\beta_1 = 3$

n	$\beta_1=3$			
	SELF	QLF	WLF	PLF
100	3.26977 0.05237	3.20438 0.00265	3.23674 0.01171	3.28658 0.01192
200	3.24543 0.02540	3.21298 0.00132	3.22912 0.00578	3.25366 0.00583
300	3.18322 0.01213	3.16730 0.00066	3.17524 0.00283	3.18723 0.00284
400	3.09412 0.00762	3.08381 0.00044	3.08896 0.00183	3.09671 0.00184
500	3.02177 0.00544	3.01421 0.00033	3.01799 0.00134	3.02366 0.00134
n	$\beta_2=6$			
	SELF	QLF	WLF	PLF

100	6.69491 0.06258	6.56101 0.00316	6.62729 0.01399	6.72933 0.01425
200	6.64508 0.03036	6.57863 0.00158	6.61168 0.00691	6.66192 0.00697
300	6.51770 0.01449	6.48511 0.00079	6.50136 0.00338	6.52590 0.00340
400	6.33527 0.00911	6.31415 0.00053	6.32470 0.00219	6.34058 0.00220
500	6.18712 0.00651	6.17165 0.00040	6.17938 0.00160	6.19100 0.00161
n	p = 0.45			
	SELF	QLF	WLF	PLF
100	0.49284 0.00413	0.48299 0.00021	0.48786 0.00092	0.49538 0.00094
200	0.48917 0.00200	0.48428 0.00010	0.48672 0.00046	0.49041 0.00046

300	0.47980 0.00096	0.47740 0.00005	0.47859 0.00022	0.48040 0.00022
400	0.46637 0.00060	0.46481 0.00003	0.46559 0.00014	0.46676 0.00014
500	0.45546 0.00043	0.45432 0.00003	0.45489 0.00011	0.45575 0.00011

Table 3. Bayes estimates and posterior risks under Jeffreys prior using β1 = 3, β2 = 6 and p = 0.30

n	$\beta_1=3$			
	SELF	QLF	WLF	PLF
100	3.26429 0.05245	3.19901 0.00265	3.23132 0.01173	3.28108 0.01194
200	3.23999 0.02544	3.20759 0.00133	3.22371 0.00579	3.24821 0.00584
300	3.17789 0.01215	3.16200 0.00066	3.16992 0.00283	3.18189 0.00285
400	3.08894 0.00763	3.07864 0.00044	3.08378 0.00183	3.09153 0.00184
500	3.01671 0.00545	3.00916 0.00033	3.01293 0.00134	3.01860 0.00135
n	$\beta_2=6$			
	SELF	QLF	WLF	PLF

100	6.55488 0.06037	6.42378 0.00305	6.48867 0.01350	6.58858 0.01374
200	6.50608 0.02928	6.44102 0.00153	6.47339 0.00667	6.52258 0.00672
300	6.38137 0.01398	6.34946 0.00076	6.36538 0.00326	6.38940 0.00328
400	6.20276 0.00878	6.18208 0.00051	6.19240 0.00211	6.20795 0.00212
500	6.05771 0.00627	6.04256 0.00038	6.05013 0.00155	6.06151 0.00155
n	$p = 0.30$			
	SELF	QLF	WLF	PLF
100	0.33081 0.00287	0.32419 0.00015	0.32747 0.00064	0.33251 0.00065
200	0.32835 0.00139	0.32506 0.00007	0.32670 0.00032	0.32918 0.00032

100	0.33081 0.00287	0.32419 0.00015	0.32747 0.00064	0.33251 0.00065
200	0.32835 0.00139	0.32506 0.00007	0.32670 0.00032	0.32918 0.00032
300	0.32205 0.00066	0.32044 0.00004	0.32125 0.00016	0.32246 0.00016
400	0.31304 0.00042	0.31200 0.00002	0.31252 0.00010	0.31330 0.00010
500	0.30572 0.00030	0.30496 0.00002	0.30534 0.00007	0.30591 0.00007

Table 4. Bayes estimates and posterior risks under Jeffreys prior using β1= 3, β2 = 6 and p = 0.45

n	$\beta_1=3$			
	SELF	QLF	WLF	PLF
100	3.24797 0.05140	3.18301 0.00260	3.21516 0.01149	3.26467 0.01170
200	3.22379 0.02493	3.19156 0.00130	3.20759 0.00568	3.23197 0.00573
300	3.16200 0.01190	3.14619 0.00065	3.15407 0.00278	3.16598 0.00279
400	3.07350 0.00748	3.06325 0.00043	3.06836 0.00180	3.07607 0.00180
500	3.00162 0.00534	2.99412 0.00032	2.99787 0.00132	3.00351 0.00132
n	$\beta_2=6$			
	SELF	QLF	WLF	PLF

100	6.68597 0.06157	6.55225 0.00311	6.61844 0.01377	6.72035 0.01402
200	6.63620 0.02987	6.56984 0.00156	6.60286 0.00680	6.65303 0.00686
300	6.50900 0.01426	6.47645 0.00078	6.49268 0.00333	6.51719 0.00334
400	6.32681 0.00896	6.30572 0.00052	6.31625 0.00215	6.33211 0.00216
500	6.17886 0.00640	6.16341 0.00039	6.17113 0.00158	6.18274 0.00158
n	p = 0.45			
	SELF	QLF	WLF	PLF
100	0.48960 0.00399	0.47981 0.00020	0.48465 0.00089	0.49212 0.00091
200	0.48596 0.00194	0.48110 0.00010	0.48351 0.00044	0.48719 0.00044
300	0.47664 0.00092	0.47426 0.00005	0.47545 0.00022	0.47724 0.00022
400	0.46330 0.00058	0.46176 0.00003	0.46253 0.00014	0.46369 0.00014
500	0.45247 0.00041	0.45133 0.00003	0.45190 0.00010	0.45275 0.00010

Table 5. Bayes estimates and posterior risks under uniform prior using $\beta_1 = 6$, $\beta_2 = 9$ and p = 0.45

n	$\beta_1=6$			
	SELF	QLF	WLF	PLF
100	6.21328 0.06556	6.15114 0.01495	6.18205 0.03122	6.22903 0.03150
200	6.11691 0.03153	6.08633 0.00747	6.10158 0.01533	6.12461 0.01540
300	6.06531 0.02062	6.04509 0.00498	6.05519 0.01013	6.07039 0.01016
400	6.07220 0.01548	6.05702 0.00374	6.06460 0.00760	6.07600 0.00762
500	5.97946 0.01200	5.96750 0.00299	5.97347 0.00599	5.98246 0.00600
n	$\beta_2=9$			
	SELF	QLF	WLF	PLF

100	9.73303 0.21487	9.63570 0.01119	9.68412 0.04891	9.75770 0.04934
200	9.54646 0.10257	9.49873 0.00560	9.52253 0.02393	9.55847 0.02403
300	9.27926 0.06445	9.24833 0.00373	9.26377 0.01549	9.28703 0.01554
400	9.06227 0.04604	9.03961 0.00280	9.05092 0.01134	9.06795 0.01137
500	9.00161 0.03631	8.98361 0.00224	8.99260 0.00901	9.00613 0.00903
n	p = 0.45			
	SELF	QLF	WLF	PLF
100	0.46979 0.00496	0.46509 0.00113	0.46743 0.00236	0.47098 0.00238
200	0.46250 0.00238	0.46019 0.00057	0.46134 0.00116	0.46309 0.00116
300	0.45860 0.00156	0.45707 0.00038	0.45784 0.00077	0.45899 0.00077
400	0.45912 0.00117	0.45797 0.00028	0.45855 0.00057	0.45941 0.00058
500	0.45211 0.00091	0.45121 0.00023	0.45166 0.00045	0.45234 0.00045

Table 6. Bayes estimates and posterior risks under Jeffreys prior using $\beta_1 = 6$, $\beta_2 = 9$ and $p = 0.45$

n	$\beta_1=6$			
	SELF	QLF	WLF	PLF
100	6.18205	6.15114	6.15114	6.19765
	0.06458	0.01487	0.03091	0.03118
200	6.10158	6.08633	6.08633	6.10924
	0.03130	0.00745	0.01525	0.01532
300	6.05519	6.04509	6.04509	6.06025
	0.02051	0.00497	0.01009	0.01012
400	6.06460	6.05702	6.05702	6.06839
	0.01542	0.00373	0.00758	0.00760
500	5.97347	5.96750	5.96750	5.97647
	0.01196	0.00299	0.00597	0.00598
n	$\beta_2=9$			
	SELF	QLF	WLF	PLF
100	9.68412	9.63570	9.63570	9.70855
	0.21164	0.01113	0.04842	0.04885

200	9.52253 0.10180	9.49873 0.00558	9.49873 0.02381	9.53449 0.02391
300	9.26377 0.06412	9.24833 0.00372	9.24833 0.01544	9.27151 0.01548
400	9.05092 0.04587	9.03961 0.00279	9.03961 0.01131	9.05659 0.01134
500	8.99260 0.03621	8.98361 0.00224	8.98361 0.00899	8.99711 0.00901
n	p = 0.45			
	SELF	QLF	WLF	PLF
100	0.46743 0.00488	0.46509 0.00112	0.46509 0.00234	0.46861 0.00236
200	0.46134 0.00237	0.46019 0.00056	0.46019 0.00115	0.46192 0.00116
300	0.45784 0.00155	0.45707 0.00038	0.45707 0.00076	0.45822 0.00077
400	0.45855 0.00117	0.45797 0.00028	0.45797 0.00057	0.45883 0.00057
500	0.45166 0.00090	0.45121 0.00023	0.45121 0.00045	0.45188 0.00045

Table 7. Bayes estimates and posterior risks under uniform prior using $\beta_1 = 6$, $\beta_2 = 9$ and p = 0.30

n	$\beta_1=6$			
	SELF	QLF	WLF	PLF
100	6.34008 0.06690	6.27668 0.01525	6.30822 0.03186	6.35615 0.03214
200	6.24175 0.03218	6.21054 0.00763	6.22610 0.01564	6.24960 0.01571
300	6.18909 0.02104	6.16846 0.00508	6.17876 0.01033	6.19428 0.01036
400	6.19612 0.01579	6.18063 0.00381	6.18836 0.00775	6.20000 0.00777
500	6.10149 0.01224	6.08929 0.00305	6.09538 0.00611	6.10455 0.00612
n	$\beta_2=9$			
	SELF	QLF	WLF	PLF

100	9.69824 0.21410	9.60126 0.01115	9.64950 0.04873	9.72282 0.04917
200	9.51233 0.10221	9.46477 0.00558	9.48849 0.02384	9.52431 0.02395
300	9.24609 0.06422	9.21527 0.00372	9.23066 0.01544	9.25383 0.01548
400	9.02987 0.04588	9.00730 0.00279	9.01857 0.01130	9.03554 0.01133
500	8.96944 0.03619	8.95150 0.00223	8.96046 0.00898	8.97393 0.00899
n	p =0.30			
	SELF	QLF	WLF	PLF
100	0.31201 0.00329	0.30890 0.00075	0.31045 0.00157	0.31281 0.00158
200	0.30718 0.00158	0.30564 0.00038	0.30641 0.00077	0.30756 0.00077
300	0.30459 0.00104	0.30357 0.00025	0.30408 0.00051	0.30484 0.00051
400	0.30493 0.00078	0.30417 0.00019	0.30455 0.00038	0.30512 0.00038
500	0.30027 0.00060	0.29967 0.00015	0.29997 0.00030	0.30042 0.00030

Table 8. Bayes estimates and posterior risks under Jeffreys prior using $\beta_1 = 6$, $\beta_2 = 9$ and $p = 0.30$

n	$\beta_1=6$			
	SELF	QLF	WLF	PLF
100	6.30822 0.06589	6.27668 0.01517	6.27668 0.03154	6.32413 0.03182
200	6.22610 0.03193	6.21054 0.00761	6.21054 0.01557	6.23392 0.01563
300	6.17876 0.02093	6.16846 0.00507	6.16846 0.01030	6.18393 0.01033
400	6.18836 0.01573	6.18063 0.00381	6.18063 0.00774	6.19224 0.00775
500	6.09538 0.01221	6.08929 0.00305	6.08929 0.00610	6.09844 0.00611
n	$\beta_2=9$			
	SELF	QLF	WLF	PLF
100	9.64950 0.21088	9.60126 0.01109	9.60126 0.04825	9.67384 0.04867

200	9.48849 0.10144	9.46477 0.00556	9.46477 0.02372	9.50041 0.02383
300	9.23066 0.06389	9.21527 0.00371	9.21527 0.01538	9.23837 0.01543
400	9.01857 0.04571	9.00730 0.00278	9.00730 0.01127	9.02422 0.01130
500	8.96046 0.03608	8.95150 0.00223	8.95150 0.00896	8.96495 0.00898
n	p = 0.30			
	SELF	QLF	WLF	PLF
100	0.30817 0.00322	0.30663 0.00074	0.30663 0.00154	0.30895 0.00155
200	0.30416 0.00156	0.30340 0.00037	0.30340 0.00076	0.30454 0.00076
300	0.30185 0.00102	0.30134 0.00025	0.30134 0.00050	0.30210 0.00050
400	0.30232 0.00077	0.30194 0.00019	0.30194 0.00038	0.30251 0.00038
500	0.29777 0.00060	0.29748 0.00015	0.29748 0.00030	0.29792 0.00030

Table 9. Bayes estimates and posterior risks under uniform prior using $\beta_1 = 10$, $\beta_2 = 12$ and $p = 0.45$

n	$\beta_1 = 10$			
	SELF	QLF	WLF	PLF
100	10.39357 0.10967	10.28963 0.02500	10.34134 0.05223	10.41992 0.05269
200	10.23237 0.05275	10.18121 0.01250	10.20673 0.02565	10.24525 0.02576
300	10.14606 0.03449	10.11224 0.00833	10.12912 0.01694	10.15455 0.01699
400	10.15757 0.02589	10.13218 0.00625	10.14486 0.01271	10.16394 0.01274
500	10.03245 0.02007	10.01238 0.00500	10.02241 0.01001	10.03748 0.01003
n	$\beta_2 = 12$			
	SELF	QLF	WLF	PLF

100	13.04696 0.28803	12.91649 0.01500	12.98140 0.06556	13.08003 0.06614
200	12.79686 0.13750	12.73288 0.00750	12.76479 0.03207	12.81297 0.03221
300	12.43869 0.08639	12.39723 0.00500	12.41792 0.02077	12.44910 0.02083
400	12.14781 0.06172	12.11744 0.00375	12.13261 0.01520	12.15543 0.01524
500	12.06651 0.04868	12.04237 0.00300	12.05443 0.01208	12.07255 0.01210
n	p = 0.45			
	SELF	QLF	WLF	PLF
100	0.47395 0.00500	0.46921 0.00114	0.47157 0.00238	0.47515 0.00240
200	0.46660 0.00241	0.46426 0.00057	0.46543 0.00117	0.46718 0.00117
300	0.46266 0.00157	0.46112 0.00038	0.46189 0.00077	0.46305 0.00077
400	0.46319 0.00118	0.46203 0.00029	0.46261 0.00058	0.46348 0.00058
500	0.45611 0.00092	0.45520 0.00023	0.45565 0.00046	0.45634 0.00046

Table 10. Bayes estimates and posterior risks under Jeffreys prior using $\beta_1 = 10$, $\beta_2 = 12$ and p = 0.45

n	$\beta_1=10$			
	SELF	QLF	WLF	PLF
100	10.34134 0.10802	10.28963 0.02488	10.28963 0.05171	10.36742 0.05216
200	10.20673 0.05235	10.18121 0.01247	10.18121 0.02552	10.21954 0.02563
300	10.12912 0.03431	10.11224 0.00832	10.11224 0.01688	10.13758 0.01693
400	10.14486 0.02579	10.13218 0.00624	10.13218 0.01268	10.15121 0.01271
500	10.02241 0.02001	10.01238 0.00500	10.01238 0.00999	10.02743 0.01001
n	$\beta_2=12$			
	SELF	QLF	WLF	PLF

100	12.98140 0.28370	12.91649 0.01493	12.91649 0.06491	13.01414 0.06548
200	12.76479 0.13647	12.73288 0.00748	12.73288 0.03191	12.78082 0.03205
300	12.41792 0.08596	12.39723 0.00499	12.39723 0.02070	12.42830 0.02076
400	12.13261 0.06149	12.11744 0.00375	12.11744 0.01517	12.14021 0.01520
500	12.05443 0.04853	12.04237 0.00300	12.04237 0.01205	12.06046 0.01208
n	p = 0.45			
	SELF	QLF	WLF	PLF
100	0.47157 0.00493	0.46921 0.00113	0.46921 0.00236	0.47275 0.00238
200	0.46543 0.00239	0.46426 0.00057	0.46426 0.00116	0.46601 0.00117
300	0.46189 0.00156	0.46112 0.00038	0.46112 0.00077	0.46227 0.00077
400	0.46261 0.00118	0.46203 0.00028	0.46203 0.00058	0.46290 0.00058
500	0.45565 0.00091	0.45520 0.00023	0.45520 0.00046	0.45588 0.00046

Table 11. Bayes estimates and posterior risks under uniform prior using $\beta_1 = 10$, $\beta_2 = 12$ and $p = 0.30$

n	$\beta_1 = 10$			
	SELF	QLF	WLF	PLF
100	10.64302	10.53659	10.58953	10.66999
	0.11230	0.02560	0.05348	0.05396
200	10.47795	10.42556	10.45169	10.49114
	0.05401	0.01280	0.02626	0.02638
300	10.38956	10.35493	10.37222	10.39826
	0.03531	0.00853	0.01734	0.01740
400	10.40135	10.37535	10.38833	10.40788
	0.02651	0.00640	0.01302	0.01305
500	10.24250	10.22202	10.23225	10.24764
	0.02055	0.00512	0.01025	0.01027
n	$\beta_2 = 12$			
	SELF	QLF	WLF	PLF

100	12.99912 0.28697	12.86913 0.01495	12.93380 0.06532	13.03207 0.06590
200	12.74994 0.13699	12.68619 0.00747	12.71799 0.03195	12.76599 0.03210
300	12.39308 0.08607	12.35177 0.00498	12.37239 0.02069	12.40345 0.02075
400	12.10327 0.06149	12.07301 0.00374	12.08812 0.01515	12.11086 0.01518
500	12.02226 0.04850	11.99822 0.00299	12.01023 0.01203	12.02829 0.01206
n	p = 0.30			
	SELF	QLF	WLF	PLF
100	0.31596 0.00333	0.31280 0.00076	0.31438 0.00159	0.31677 0.00160
200	0.31106 0.00160	0.30951 0.00038	0.31028 0.00078	0.31146 0.00078
300	0.30844 0.00105	0.30741 0.00025	0.30793 0.00051	0.30870 0.00052
400	0.30879 0.00079	0.30802 0.00019	0.30840 0.00039	0.30898 0.00039
500	0.30407 0.00061	0.30347 0.00015	0.30377 0.00030	0.30423 0.00030

Table 12. Bayes estimates and posterior risks under Jeffreys prior using $\beta_1 = 10$, $\beta_2 = 12$ and $p = 0.30$

n	$\beta_1=10$			
	SELF	QLF	WLF	PLF
100	10.58953	10.53659	10.53659	10.61624
	0.11062	0.02547	0.05295	0.05342
200	10.45169	10.42556	10.42556	10.46481
	0.05361	0.01277	0.02613	0.02624
300	10.37222	10.35493	10.35493	10.38088
	0.03514	0.00852	0.01729	0.01734
400	10.38833	10.37535	10.37535	10.39484
	0.02641	0.00639	0.01299	0.01301
500	10.23225	10.22202	10.22202	10.23737
	0.02049	0.00511	0.01023	0.01025
n	$\beta_2=12$			
	SELF	QLF	WLF	PLF
100	12.93380	12.86913	12.86913	12.96642
	0.28266	0.01487	0.06467	0.06524

200	12.71799 0.13597	12.68619 0.00745	12.68619 0.03179	12.73395 0.03193
300	12.37239 0.08564	12.35177 0.00497	12.35177 0.02062	12.38273 0.02068
400	12.08812 0.06126	12.07301 0.00373	12.07301 0.01511	12.09569 0.01514
500	12.01023 0.04836	11.99822 0.00299	11.99822 0.01201	12.01624 0.01203
n	p = 0.30			
	SELF	QLF	WLF	PLF
100	0.31024 0.00324	0.30869 0.00075	0.30869 0.00155	0.31102 0.00156
200	0.30620 0.00157	0.30544 0.00037	0.30544 0.00077	0.30659 0.00077
300	0.30387 0.00103	0.30337 0.00025	0.30337 0.00051	0.30413 0.00051
400	0.30435 0.00077	0.30397 0.00019	0.30397 0.00038	0.30454 0.00038
500	0.29977 0.00060	0.29947 0.00015	0.29947 0.00030	0.29992 0.00030

Table 13. 95% posterior predictive intervals for $\beta_1 = 3, \beta_2 = 6$, and $p = 0.30$

n	Uniform		
	Lower Limit	Upper Limit	Difference
50	2.95364	16.40910	13.45546
100	2.93165	16.28695	13.35530
200	2.87545	15.97475	13.09929
300	2.79497	15.52763	12.73266
400	2.72961	15.16452	12.43490
500	2.71134	15.06302	12.35168
n	Jeffreys		
	Lower Limit	Upper Limit	Difference
50	2.65827	15.58864	12.93037
100	2.63849	15.47261	12.83412
200	2.58791	15.17601	12.58810
300	2.51548	14.75125	12.23577
400	2.45665	14.40629	11.94964
500	2.44021	14.30987	11.86966

From tables 13-16 it can be assessed that posterior predictions tend to be more accurate in larger samples. Increasing values of actual and weight parameter imposes negative impact on the performance of the predictions. It is interesting to note that the posterior predictive intervals are shorter in case of Jeffreys prior. This simply indicates the supremacy of Jeffreys prior over uniform prior. Anyhow, the results under the posterior predictive intervals are in accordance with the corresponding point estimation.

Table 14. 95% posterior predictive intervals for $\beta_1 = 3$, $\beta_2 = 6$, and $p = 0.45$

n	Uniform		
	Lower Limit	Upper Limit	Difference
50	3.01271	16.73728	13.72457
100	2.99028	16.61269	13.62241
200	2.93296	16.29424	13.36128
300	2.85087	15.83818	12.98731
400	2.78421	15.46781	12.68360
500	2.76557	15.36428	12.59871
n	Jeffreys		
	Lower Limit	Upper Limit	Difference
50	2.71144	15.90042	13.18898
100	2.69126	15.78206	13.09080
200	2.63967	15.47953	12.83986
300	2.56579	15.04627	12.48049
400	2.50578	14.69442	12.18863
500	2.48901	14.59607	12.10705

Table 15. 95% posterior predictive intervals for $\beta_1 = 6$, $\beta_2 = 9$, and $p = 0.30$

n	Uniform		
	Lower Limit	Upper Limit	Difference
50	4.43046	22.97274	18.54228
100	4.39748	22.80173	18.40426
200	4.31318	22.36465	18.05147
300	4.19246	21.73868	17.54622
400	4.09442	21.23032	17.13590
500	4.06702	21.08823	17.02121
n	Jeffreys		
	Lower Limit	Upper Limit	Difference
50	3.98741	21.82410	17.83669
100	3.95773	21.66165	17.70392
200	3.88186	21.24642	17.36455
300	3.77321	20.65175	16.87853
400	3.68498	20.16881	16.48383
500	3.66031	20.03382	16.37350

Table 16. 95% posterior predictive intervals for $\beta_1 = 3, \beta_2 = 6$, and $p = 0.30$

n	Uniform		
	Lower Limit	Upper Limit	Difference
50	4.87350	26.64838	21.77488
100	4.83723	26.45001	21.61279
200	4.74450	25.94299	21.19849
300	4.61171	25.21687	20.60516
400	4.50386	24.62717	20.12331
500	4.47372	24.46235	19.98863
n	Jeffreys		
	Lower Limit	Upper Limit	Difference
50	4.38615	25.31596	20.92981
100	4.35350	25.12751	20.77401
200	4.27005	24.64584	20.37579
300	4.15054	23.95602	19.80549
400	4.05348	23.39582	19.34234
500	4.02635	23.23923	19.21288

7. ANALYSIS UNDER REAL LIFE DATA

The real life data (following Gumbel distribution) regarding monthly wind speed in Cameron Highland from year 2004-2006 presented by Zaharimi et al.[18] is used to illustrate the applicability of the results obtained in previous sections.

Table 17. Bayes estimates and risks for real life data

Prior	β_1			
	SELF	QLF	WLF	PLF
Uniform	3.18955 0.30934	3.00000 0.02984	3.09422 0.09533	3.23672 0.09435
Jeffreys	3.09422 0.30098	2.33867 0.03109	3.00000 0.09422	3.14152 0.09460
	β_2			
	SELF	QLF	WLF	PLF
Uniform	3.58011 0.34722	3.36735 0.03350	3.47311 0.10700	3.63306 0.10590
Jeffreys	3.47311 0.33784	2.62504 0.03490	3.36735 0.10576	3.52620 0.10619
	$p = 0.30$			
	SELF	QLF	WLF	PLF
Uniform	0.32546 0.03157	0.30612 0.00305	0.31574 0.00973	0.33028 0.00963
Jeffreys	0.31574 0.03071	0.23864 0.00317	0.30612 0.00961	0.32056 0.00965

Table 18. Bayes estimates and risks for real life data

Prior	β_1			
	SELF	QLF	WLF	PLF
Uniform	3.28205 0.31831	3.08700 0.03071	3.18396 0.09809	3.33059 0.09708
Jeffreys	3.18396 0.30971	2.40650 0.03199	3.08700 0.09696	3.23263 0.09735
	β_2			
	SELF	QLF	WLF	PLF
Uniform	3.30346 0.32039	3.10714 0.03091	3.20473 0.09873	3.35232 0.09772
Jeffreys	3.20473 0.31173	2.42220 0.03220	3.10714 0.09759	3.25372 0.09798
	p = 0.45			
	SELF	QLF	WLF	PLF
Uniform	0.47192 0.04577	0.44388 0.00442	0.45782 0.01410	0.47890 0.01396
Jeffreys	0.45782 0.04453	0.34603 0.00460	0.44388 0.01394	0.46482 0.01400

The real life data replicated the patterns observed in the simulation study. The minimum amount of risks has been observed for the estimates under the assumption of Jeffreys prior using quadratic loss function.

8. CONCLUSIONS

The purpose of the article is to find out the appropriate combination of prior distribution and loss function to estimate the parameters of two-component mixture of Gumbel type II distribution. The parameters have been estimated under the assumption of two non-informative priors and four loss functions (symmetric and asymmetric). The posterior predictive intervals have also been evaluated. From the findings of the study it can be concluded that in order to estimate the said parameters, the use of Jeffreys prior and quadratic loss function can be preferred.

REFERENCES

1. R. Chechile, "Bayesian analysis of Gumbel distributed data", Communications in Statistics - Theory and Methods, vol.30, no.3, pp. 211-224, 2001.
2. G. Corsini, F. Gini, M. V. Gerco, et al. "Cramer-Rao bounds and estimation of the parameters of the Gumbel distribution", IEEE, vol.31, no.3, pp. 1202-1204, 2002.
3. A. Mousa, "Bayesian estimation, prediction and characterization for the Gumbel model based on records", Statistics, vol.36, no.1, pp. 58-69, 2002.
4. D. Koutsoyiannis, and G. Baloutsos, "Analysis of a long record of annual maximum rainfall in Athens, Greece, and design rainfall inferences", Natural Hazards, vol.22, no.1, pp. 29-48, 2000.
5. Rasmussen, P. F., and Gautam, N. "Alternative PWM-estimators of the Gumbel distribution. Journal of Hydrology", vol.280, no.1-4, pp. 265-271, 2003.
6. Malinowka, and D. Szynal, "On characterization of certain distributions of kth lower (upper) record values", Applied Mathematics and Computation, vol.202, no.1, pp. 338-347, 2008.
7. S. Nadarajah, and S. Kotz, "The beta Gumbel distribution", Math. Probab. Eng., vol.10, 323-332, 2004.
8. Miladinovic, and P. C. Tsokos, "Sensitivity of the Bayesian reliability estimates for the modified Gumbel failure model", International Journal of Reliability, Quality and Safety Engineering (IJRQSE), vol.16(40), pp. 331-341, 2009.
9. [9] Y. Park, S. Sheetlin, and L. J. Spouge, "Estimating the Gumbel scale parameter for local alignment of random sequences by importance sampling with stopping times", Ann. Statist., vol.37, no.6A, pp. 3697-3714, 2009.
10. H. J. Heo, and D. J. Salas, "Estimation of quantiles and confidence intervals for the log-Gumbel distribution", Stochastic Hydrology and Hydraulics, vol.10, no.3, pp. 187-207, 2010.

11. E. M. Thompson, J. B. Hewlett, and R. M. Vogel, "The Gumbel hypothesis test for left censored observations using regional earthquake records as an example", Nat. Hazards Earth Syst. Sci., vol.11, pp. 115-126, 2011.

12. M. Saleem, and M. Aslam, "Bayesian Analysis of the Two Component Mixture of the Rayleigh Dist. With the Uniform and the Jeffreys Priors", J. of Applied Statistical Science, Vol.16, no.4, pp.105-113, 2008.

13. M. Saleem, M. Aslam, and P. Economou, "On the Bayesian analysis of the mixture of power function distribution using the complete and the censored sample", Journal of Applied Statistics, vol.37, no.1, pp. 25-40, 2010.

14. M. Y. Majeed, and M. Aslam, "Bayesian analysis of the two component mixture of inverted exponential distribution under quadratic loss functions", International Journal of Physical Sciences, vol.7, no.9, pp. 1424-1434, 2012.

15. S. M. A. Kazmi, M. Aslam, and S. Ali "On the Bayesian estimation for two component mixture of maxwell distribution, Assuming Type I Censored Data", International Journal of Applied Science and Technology (IJAST), vol.2, no.1, pp. 197-218, 2012.

16. P. S. Laplace, "Theorie analytique des probabilities", Veuve Courcier Paris, 1812.

17. H. Jeffreys, "Theory of Probability", 3rd edn. Oxford University Press, pp. 432, 1961.

18. A. Zaharimi, S. Najid, and A. Mahir, "Analyzing Malaysian wind speed data using statistical distribution", Proceedings of the 4th IASME / WSEAS Inter. Conf. on Energy & Environment, Malysia, pp. 363-370, 2009.

Index

M

P

R

S